Feathers
on the
Frontier

A Portrait Of Pioneer Poultry Keeping

KATHRYN STEVENSON

FEATHERS ON THE FRONTIER
A Portrait Of Pioneer Poultry Keeping

Published by Kathryn Stevenson, Alberta, Canada

ISBN:
    Paperback   978-1-77354-710-7
        ebook   978-1-77354-717-6

Publication assistance by

PAGEMASTER
PUBLISHING
PageMaster.ca

Acknowledgement

The regions described in this book were the traditional lands and territories of the Indigenous Peoples. What became known as the Province of Alberta in 1905 is the ancestral home to many Indigenous Peoples in the Treaty 6, 7, and 8 regions. Alberta is also within the historical Northwest Métis Homeland. This acknowledgment is a statement of gratitude to the many First Nations, Métis, and Inuit people who have lived in and cared for these lands for generations.

Contents

Preface

I feel privileged to have had the experience of growing up on a
small mixed farm in central Alberta. At the time, I did not think
my childhood was anything special. Ours was a typical family farm
of the '70s where grain was grown, cattle were raised, and a few
pigs were kept. Kittens were born in hay lofts, calves nuzzled and
bunted buckets of rich milk, and chickens were quick to tidy up
any oats, barley, or wheat scattered around the grain bins.

Chickens were part and parcel of farm life. Each year, my
mom raised a batch of broilers to fill the freezer. I did not find the
white Cornish-cross chickens attractive or enjoyable, but they were
a necessary food source. The fun flock was a menagerie of hens
and roosters that had the run of the farm. The hens were not big
enough to be all that useful, and not small enough to be proper
bantams...they were just run-of-the-mill farmyard chickens. They
roamed freely from early spring until late fall, raising their young
and often roosting high in the trees at night. When the weather
turned cold, they were captured and confined to the coop for the
winter. There were sometimes squabbles among the feisty roosters,
but mostly they lived in harmony.

A couple of decades later, I was fresh out of college and newly married. My husband and I purchased our own little piece of paradise near Buffalo Lake, Alberta, a stone's throw from his family's original homestead. An old chicken house sat on the property, long neglected and mostly hidden by brush. It had probably been built somewhere in the 40s or early 50s, and while it was still sturdy, it needed some care. We gave it a good cleaning, patched up the roof, replaced the windows, and made it service-able. The following spring, I put in an order for 25 broiler chicks from the nearest hatchery. Besides the usual choices of white meat chickens or hybrid layers, the hatchery carried a selection of "heritage breeds," and their enticing photos in the catalogue were impossible to resist.

The heritage chick package introduced me to some historic breeds I'd only seen in pictures previously. They were fun to raise, and I soon learned that each breed had a fascinating history! I began to study everything I could about the traditional farm chickens—the kinds commonly found throughout Canada at the turn of the century. I was stunned to discover that many of the old breeds were disappearing. Granted, some were still available through hatcheries, but since they were no longer being bred and used for their intended purposes, most hatchery lines had regressed to mere shadows of their former glory. It seemed tragic to think that such a cherished and significant phase of agricultur-al history could be slipping quietly into obscurity. The continued existence of authentic, historic poultry strains depended on a few dedicated breeders and enthusiasts. That shocking realization spurred my transition from chicken keeper to breeder, conserva-tionist, and exhibitor. It was the start of a wonderful, rewarding hobby that I still enjoy today.

Still, I had so many questions about poultry keeping during the pioneer era. How did settlers manage their flocks in such a rugged environment? Did they find chicken-keeping profitable?

Which breeds were popular on the Canadian prairies? Were incubators frequently used? When were poultry exhibitions first held in the Western provinces? Online searches revealed some clues, but the information mainly centered around Ontario's poultry scene, a much more advanced poultry culture in the early 1900s. In a fortunate stroke of serendipity, many answers lie in an antique trunk, just waiting to be discovered.

It was my husband's grandmother's trunk. Mr. & Mrs. H. Stevenson were among the first homesteaders to arrive and settle on the west shores of Buffalo Lake, Alberta, some 120 years ago. Along with some letters and journals, they had collected and carefully stored away a dozen or so agricultural newspapers dating back as far as 1908. The trunk came into our possession along with its precious contents. The stiff, yellowed pages of *The Farmer's Advocate and Home Journal*, and *The Farm and Ranch Review* were a vivid and fascinating window into the past. The discovery of those materials spurred further research, and what has emerged is a collection of the common practices and personal experiences of raising chickens during the tumultuous years of 1900 to 1920. Those men and women had to be tough, spirited, and adaptable to survive, and those same, admirable qualities were integral features of the fowl they raised.

Log Brooder House on the Leon and Carol Young Family Farm, Bashaw

Introduction

May, 1903

The horse-drawn covered wagon creaked and swayed, bumping along the casually winding trail. A man in his early twenties held the reins loosely; his dutiful team needed little encouragement to follow the well-worn path. He adjusted the brim of his hat to better shade his eyes from the sun's slanting rays while the woman next to him shifted to find a more comfortable position on the hard, wooden seat.

Their adventure began in Liverpool in the spring of 1903. The newlyweds had boarded a ship bound for St. John, Canada, with a teary farewell to their parents and siblings. The past weeks have been a blur for Jonathan and Isabelle, beginning with the transatlantic crossing and then the trip west by rail. The train brought them to the bustling town of Lacombe, where they stepped from their passenger car into a pandemonium of other settlers and townspeople. Eventually, the crowds dissipated, and the couple found their way to a boarding house. They spent two weeks in Lacombe, restoring their spirits and preparing for the final leg of their journey.

Their destination lay at the end of a well-established route called Buffalo Lake Trail. Mile after mile, the trail wandered and curved through gently rolling hills and into muddy, low-lying areas that delayed their progress. The wagon groaned under the weight of the personal belongings they brought from England, plus the equipment and supplies they had purchased in Lacombe. A crate tied to the rear of the wagon, while not especially weighty, held a particularly prized purchase: one cockerel and six pullets.

Their third day on the trail brought the weary travellers to the village of Lamerton. The small but lively community seemed a sensible place to rest for the night, and the couple considered setting up camp. But they were so close now; their own quarter-section of land called to them. Isabelle and Jonathan decided to press on.

Buffalo Lake Trail began to dwindle and was soon reduced to wheel ruts cutting through rippling native grasses, prairie sage, and wildflowers. At long last, Jonathan halted the horses and climbed down from the wagon. A great expanse of water was visible in the distance—Buffalo Lake. The sharp, camphor fragrance of sage enveloped them, mingling with the scent of lakeshore on the breeze. A robin chose that moment to commence his evening song in a clear, lilting solo that rose above the quiet rustle of leaves and grass.

Jonathan glanced again at the survey marker he had located—a simple wooden stake marking the end of a long, arduous journey and the beginning of a new life. Isabel appeared at his side, a sparkle in her eyes and a smile on her lips. They were home.

⚜ ⚜ ⚜ ⚜ ⚜

This is not an actual account of a real couple's journey; it depicts a moment in time that countless Europeans might have experienced during Western Canada's immigration boom of the early 1900s.

Wave after wave of settlers descended on the prairies with the force of a tsunami, quickly snapping up the fertile land for farms and ranches. And the Canadian government was ready. The land had been surveyed into quarter-section homesteads (160-acre parcels) awaiting registration. The only cost for the land was a $10.00 application fee, but a few conditions were attached. Before gaining full ownership, homesteaders would need to "prove up". This meant complying with some requirements such as erecting a dwelling, breaking ground, growing crops, or running cattle.

It is a unique, fascinating chapter in Canadian history that defines and illustrates the strength of the human spirit. Newcomers to this land arrived with dreams and aspirations but were often unprepared for the rigors and trials that awaited them. Those from a farming background had distinct advantages, but all who succeeded did so through hard physical labour and perseverance.

Feathers on the Frontier

As the wild vistas of the West became dotted with farmsteads, small flocks of chickens began to occupy them. Turkeys, domestic geese, and ducks, while not as widespread as chickens, became valued farm stock to those who raised them. Not every farmer raised purebreds, but those who did kept the hardy, traditional fowl we now affectionately refer to as heritage poultry. While no longer abundant, these breeds still have a presence in our modern world...in farm flocks, on acreages, and even in backyards within cities and towns where urban hens are permitted.

Poultry exhibitors still compete for ribbons, prizes, and bragging rights at fairs and exhibitions, just as they did before the turn of the century. Spectators at these shows can admire a stunning array of plumage colours and feather patterns—the same patterns that adorned the hens and roosters of the past, housed in humble log huts with mud-chinked walls.

And the perpetuity of the historic breeds runs much deeper than feathers. They still charm us with curious and quirky behaviors in all the same ways they entertained and delighted their pioneer caregivers. We can still cheer for the nimble pullet who manages to capture a grasshopper, then dashes to evade her sharp-eyed pen-mates who would snatch it away in an instant. Hens still delight in creating dry, sandy hollows for dustbathing. They lounge and stretch in the warmth of midday, then rise, shaking the dirt from their bodies like a dog shedding water.

Chickens still rake the earth in their incessant search for seeds, shoots, and insects, performing that distinctive scratch-dance perfected by their distant ancestors. Present-day farmyards are still filled with the exuberant crow of a rooster announcing the dawn, but seasoned chicken keepers know that sunrise is only the beginning of his random and frequent announcements that will continue throughout the day. Broody hens protect their nest with raised hackles, a menacing whine, and often a quick, hard peck to the hand of a would-be egg-gatherer. And newly hatched chicks, some camouflaged in downy stripes of chocolate and gold, still rush to find safety and warmth under their mother's protective wings.

Historical poultry-keeping is not a common topic of study, but it is a worthy one. It offers a captivating glimpse into the everyday lives of those brave and ambitious souls who settled Canada's West at the dawn of a new century.

Chapter 1

a New Life in a New Land

September, 1903

Isabelle was seated on a little wooden stool just outside the front entrance of their house, up to her elbows in soapy water. She paused from her laundry chores, lifting her face to the mellow warmth of the late-September sun. Her chickens had gathered nearby, and she watched them with quiet pleasure. They were free to roam during the day but spent much of their time in Isabelle's company, picking through the loose, sandy soil surrounding the house. On rainy days, they preferred to stay dry and comfortable inside their coop. Isabelle found their simple ways and familiar routines comforting.

The chicken house certainly wasn't fancy, but it served its purpose. Jonathan and Isabelle had worked together to build a sturdy structure to keep the flock safe. Areas of their property were densely wooded, and poplar trees grew straight and tall. After felling the trees, Jonathan had made quick work of notching and stacking the logs to form the four walls. While he worked on the roof, Isabelle chinked the walls with a clay and mud mixture. There was even one glass window. Isabelle had insisted a south-facing window was necessary to let in the sun's light and warmth.

Their bachelor neighbour, Ian, had shown Isabelle how to make the mud mixture and fill the gaps between the logs. Having Ian nearby had been a lucky break. He had helped Jonathan build their house, working long hours nearly every day until it was complete. Ian was happy for the company—and the hearty meals—and he accepted only a modest wage for his work. It certainly wasn't a grand home, but it was well-built, and Isabelle was relieved their first winter in Canada would be spent in a cozy, wood-frame house. Ian's log shack was not much bigger than her chicken coop.

           ❧ ❧ ❧ ❧ ❧

The farmers and ranchers who settled western Canada came heavily laden—both with supplies and determination—but they faced a daunting task. Building a home and a farm from a bare patch of land was a massive undertaking. Some arrived with crates of chickens or livestock in tow, and an immediate concern was to shelter them... their own house might have to wait. Families often lived in tents for months while permanent shelters were being built. Others improvised with a quick, temporary lodging such as a small sod or log shack. Some even started with a dual-purpose building: living quarters in one end, barn in the other, all under one roof.

Real, first-hand experiences are the most powerful and poignant way to illustrate the pioneer lifestyle. Let's step away from fiction and become acquainted with some actual families and their true, first-hand accounts of setting up a farmstead and caring for their poultry.

[Image 101: "Mrs. Chapman washing clothes, Endiang, Alberta.", 1916, (CU1127198) by Unknown. Courtesy of Glenbow Library and Archives Collection, Libraries and Cultural Resources Digital Collections, University of Calgary.]

Newcomers to the North

John and Mary Walton came to Alberta as homesteaders in 1910. John was managing a grocery store in Toronto when he first heard about the Peace Country of Northern Alberta. Acquaintances had left Ontario to settle in that region, and their letters were enticing. With their interest sufficiently piqued, the Waltons decided to follow suit. The couple arrived in Edmonton in mid-February, purchased oxen, sleighs, supplies, and equipment, and were ready to hit the trail by March 1st. Their route took them to Athabasca Landing, across the frozen Lesser Slave Lake, and eventually to their property near Beaverlodge, Alberta.

Completing such a journey would have been a tremendous feat in itself, but looming ahead was the grueling work of clearing land and creating a farm from scratch. At first, John and Mary's only structure was dual-purpose; the family at one end, livestock at the other, and a wall in between. Mrs. Walton admitted to living in constant fear of fire. They used flares to light their home, and she worried that if the building ever caught fire, all would likely be lost: their home, belongings, and the livestock.

Fires were indeed a top concern for pioneer families, along with about a hundred other possible calamities. But like so many others, the Waltons adjusted to frontier life. They started raising 'huge' flocks of turkeys. Coyotes were always on the lookout for an easy meal, so the birds were safely tucked away at night, and during the day, they roamed under the protection of herders: John and Mary's daughters. The girls did double duty during their long hours of following the flock. They carried their knitting supplies and used the time constructively to create socks, mittens, and sweaters. The Walton family might have had plenty to worry about, but warm winter clothing was not one of them.

Children learned the basics of poultry care from a young age. Feeding the farmyard fowl was an activity suitable for all ages,

Josephine Bedingfield on the family ranch located west of High River, Alberta, 1915

and history is peppered with photos of young children scattering grain for the family flock. Beyond its practicality, attending to the chickens was a source of entertainment—many children delighted in making pets of chicks or ducklings.

Another family arrived in the same region two years later— the McNaughts. Sisters, Betty and Isabel, travelled across Canada to the wilds of Northern Alberta, and their exploits have been expertly recorded and preserved. Their vivid memories offer a richly detailed account of this remarkable time in Alberta's history.

Runaway Rooster

Euphemia (Betty) McNaught and her sister, Isabel, were children when they left their comfortable home in Glenmorris, Ontario, in 1912. Their parents, Charles and Eliza, rented a boxcar, and although the space was shared with another family, there was room enough to transport their two Percheron horses, three cows, and nine chickens to Alberta. The train brought the family, their belongings, and the livestock as far as Edson, but their land was located near Beaverlodge—a six-week wagon journey farther north.

Betty recalled her father stepping down from the wagon, finding the little iron corner post, and declaring, "If I read this right, we are home." They found a sunny, open area near a lake to set up camp, pitch the tent, and build a temporary corral. But, while the fence kept the cows contained, the rooster was a rambler.

Charles had a lovely flock of Blue Andalusian hens, and just before leaving Ontario, he had been gifted a Red Game rooster. The first day on their new property began with the discovery that the rooster had abandoned camp. Betty's parents set off to track him down. After a long trek around the lake, they finally caught up with the rogue rooster, captured him, and carried him back. Eliza took it all in stride. "It didn't daunt her," Betty recalled. "She was very active and was ready to tackle anything. She was tall and wiry, and very slim, and probably camping and walking and all those things were not much of a hardship for her."

Game birds and Andalusians were not the usual, run-of-the-mill homestead chickens, but they sure would have been eye-catching. Game cocks are tall and strong, with snug-fitting feathers. One 19th-century poultry guide compared the long, lean head of a Game cock to that of a greyhound. Although the book did not make further comparisons between the two species, it seems—this rooster at least—shared the greyhound's enthusiasm for physical activity.

Blue Andalusian hens are best known for their astonishing plumage. Actually, both males and females are adorned with slate-blue feathers that lie smoothly on their bodies, each edged with a fine black border. It is as if a steady hand inked the perimeter of every feather with razor-sharp precision. They have matching steel gray legs, white oval earlobes, and, for a bold finishing touch, deep-red combs and wattles. Jaunty and elegant, Andalusians strut and flaunt their stylish attire, fully aware of their elegance.

Rather showy birds for a rudimentary farmyard, perhaps, but not entirely impractical. Andalusian hens are prolific egg layers and less refined than their more familiar Leghorn cousins. They are built heavier and are wonderfully adept at foraging. But practicality aside, maybe the McNaughts took comfort in watching their fancy fowl scratching about the farmyard, their regal presence in sharp contrast to the backdrop of forested foothills. Their flock was a living connection to the comforts they had left behind, and possibly a vision of optimism for the future.

This tale of the roaming rooster might have seemed inconsequential at the time, hardly worth taking note of. After all, the very act of homesteading was a superhuman feat of survival, and trekking around a lake in pursuit of a rooster probably did not seem like much in the big scheme. But it was a significant enough event to be remembered by the McNaught children, and it became a tale worth telling. This serves as a reminder that personal stories, even light-hearted anecdotes, all combine to form a rich tapestry of history that adds depth and context to our understanding of the past.

Settling the South

From around 1896 to 1914, Alberta's south was a hotbed of settlement activity. Men and women from different countries and social backgrounds came to the prairies, ordinary people who faced extraordinary circumstances. Success was not a guarantee. A CBC news article reported that only three out of every five homesteaders managed to secure their homestead patent. But many did succeed. They cultivated the land, supported each other, and built communities. Some individuals' accomplishments went far beyond

[Image 103 "Brealy Ranch, Big Hill Spring, Alberta.", [ca. 1900-1907], (CU1157447) by Unknown. Courtesy of Glenbow Library and Archives Collection, Libraries and Cultural Resources Digital Collections, University of Calgary.]

the ordinary. Mabel Barker and H.B. Biggs were two such individuals whose actions created inspirational and long-lasting legacies.

The Remarkable Mabel Barker

Mabel Barker was a 17-year-old when she came from Ontario to the Calgary, Alberta area. Her uncle had purchased a large piece of land a year earlier, in 1906. So, by the time Mabel's family arrived on a tourist train, a good, solid house had been constructed and made ready for them. Mabel became friends with a fellow named Will Barker and his sister, who lived in the neighbourhood. She ended up marrying Will, and the couple got down to the business of farming, with one distinct difference from nearly all other prairie homesteads of the time. For the Barkers, raising chickens was more than just a sideline; it was a carefully managed and profitable branch of their farming operations.

Mabel was no novice when it came to chicken farming. She had helped her mother raise poultry back in Ontario, so she knew exactly how to go about it. Each spring, she bought 500 chicks,

reared them in a brooder house with a huge chicken yard, and successfully established a business selling cockerels, pullets, and eggs.

Mabel's breed of choice was the Orpington. She also tried
raising Leghorns for a while. "I did not like them. I like the
heavy fowls," she stated in an interview. Mabel understood that
the lighter breeds were considered more prolific layers, but the
cockerels were not marketable. She preferred the pragmatic,
dual-purpose Orpingtons, as she could make a tidy profit selling
dressed cockerels.

Will built "covered nests" for the laying hens—small units
placed in rows around the yard during the summer, each with a
built-in perch. "You lifted the lid to get your eggs out. It kept them
really very healthy", Mabel explained, referring to the hens. "There
is nothing better for chickens than open air and clean grass!"

Mabel had many local customers who came to the farm
regularly to buy eggs. She carefully cleaned, weighed, and sorted
her eggs, selling them according to four different grades. When
she had an abundance, Mabel would box them up and take them
to the Egg Producers in town.

Mabel's son later took over the job of raising the cockerels,
using fattening pens to grow them out. "Of course, we used to raise
them until they would be about six pounds, five and six pounds.
And then sell them."

The Barker's main henhouse could comfortably hold 130
chickens. Mabel made sure she had reduced her flock to the desired
number of birds by the time winter rolled around. She felt it was
important that they were not overcrowded while confined indoors
during the cold months.

Although wonderfully detailed, the interview does not tell us
where she was able to buy 500 chicks so early in the 1900s. Reliable
rail service likely made it possible to have chicks shipped from
larger, more well-established centres such as Winnipeg, and by
1917, small hatcheries were up and running in Alberta.

Mabel was one of Alberta's first commercial producers. Her personal account gives us a broader understanding of the frontier poultry woman, even reshaping the way pioneer chicken-keeping has generally been viewed—as only small, mixed-breed flocks kept for personal consumption.

Mabel Barker was more than an accomplished poultry keeper and businesswoman; she was a community pillar. Active in the Red Cross and United Farm Women of Alberta, Mabel was an advocate for the women's suffrage movement, which fought to address fundamental issues of equity and justice. She was inducted into the Alberta Agriculture Hall of Fame in 1975.

The Biggs Legacy

Hugh Beynon Hesketh-Biggs emigrated from India to the Rosebud Valley, west of Drumheller, Alberta, in 1893. There, he bought Springfield Ranche. It was, perhaps, a grandiose name for little more than a sod-roofed log cabin, but it launched his dream of ranching. He got to work, building a new log house and importing cattle. In 1906, H. B. married Mabel Florence James. Mabel had been a nurse in Guernsey, England, and immigrated to Canada in 1903. Together, the Biggs had four daughters.

H. B. Biggs was an avid photographer. Through the lens of his camera, he captured warm family moments and everyday ranch life, all set against the beautiful imagery of southern Alberta. Their flocks of Barred Rocks and Narragansett turkeys were even deemed worthy subjects, and one especially poignant photo features Mabel surrounded by her chickens. While feeding them, one youngster hopped onto her lap, eliciting a reaction of unmistakable delight. During a time when photography was still breaking free from the 19th-century norms of staid, solemn expressions, a farm woman's joyful interaction with her feathered friends was preserved forever.

[Image 104: "Mabel Biggs feeding her chickens, Springfield Ranch, Beynon, Alberta.", [ca. 1910], (CU170662) by Biggs, H. B.. Courtesy of Glenbow Library and Archives Collection, Libraries and Cultural Resources Digital Collections, University of Calgary.]

A rail line came through the region and dissected Springfield Ranche in the process. (The spelling was later changed to Springfield Ranch.) A rail siding was established and named Beynon, and soon a village of the same name sprang up with a general store, blacksmith shop, school, and post office. By the late 1920s, Beynon was a small community, all situated within Springfield Ranch property. It even boasted two grain elevators, corrals for shipping cattle, and a 'section house' for the rail line maintenance crews.

Beynon eventually suffered the fate of so many prairie villages. People left for larger centres, and the businesses closed. But, while the village no longer exists, the legacy of Springfield Ranch and the Biggs family continues. Their land was deemed ecologically sensitive, and the Beynon family took measures to protect it from development, preserving its natural state for the enjoyment of future generations.

Coop-building Basics

As settlers continued to arrive and farmland was grabbed up, a burgeoning agricultural industry took root and began to grow. Wheat was the leading export product, but homesteaders knew the risk of putting all their eggs in one basket. Most hedged their bets against crop failure through mixed farming (raising both livestock and grain crops). Despite the achievements of a few forward-thinking pioneers like Mrs. Barker, poultry was not yet an established farm commodity. It was, however, a food security safety net, and nearly every farmyard had a few chickens scattered about. Some farmers raised turkeys, ducks, or geese to supplement their income.

As Alberta's population surged, so did the demand for poultry products—eggs, meat, and live breeding stock, and local production fell far short of meeting these needs. Guidance was needed to encourage landowners to raise poultry and give them tools for success. The large population of immigrants from milder climates needed specific instructions to prepare for the harsh winter season and give their birds the best chance of survival. In response, the Alberta government launched an expansive educational campaign.

Alberta Agriculture published a series of bulletins called *Successful Poultry Raising*, complete and comprehensive guides with specific relevance to the province's conditions and climate. A section on housing provided dimensions, building materials, and illustrations for building and outfitting chicken coops. Solid construction, insulation, and ventilation were the three key considerations in coop design.

Hen houses were not normally heated in the winter, nor did the experts advise it, stating that appropriate planning and construction would provide adequate shelter for overwintering healthy adult birds. "Proper accommodation is an essential in profitable poultry raising that cannot be overlooked. While buildings may be of as elaborate construction as the purse and fancy of the owner dictate, still houses constructed of log, sod, baled straw, or even straw packed between a framework are capable of giving satisfactory results when proper care is given to location, light, ventilation and sanitary conditions."

A ventilation system was mandatory for adequate air circulation without draftiness. This could be achieved by allowing outside air to enter through a small, burlap-covered window, circulate slowly through the coop, and then escape through a ceiling vent into the building's straw-insulated apex. Safe ventilation prevented the accumulation of frost in the coop, which made the inside temperature even colder. It also melted during a warming spell, resulting in dampness that could jeopardize the health of the chickens.

The chicken house should include an attached scratching shed. The main coop was where the hens slept and laid their eggs, but a covered, protected annex provided space for the flock to mingle and scratch for grain. Exercise was thought to keep chickens in a healthy condition and better able to survive the rigours of winter. Poultry stockmen were advised to allow for growth when constructing their chicken house, arranging their building "so that extensions may be easily made as required." This sage advice remains relevant throughout the ages!

Referencing the poultry yard, the manual declared that a gentle slope would provide natural drainage and be "suitable for growing grass, alfalfa, or other succulent foods suitable for poultry." An enclosed chicken run was not a regular feature on farmsteads, and most domestic birds remained free to range during the day. The manual provided one brief statement on the subject of fencing: "Installing a wire fence helps protect flocks."

Protection from Predators

Since domestic birds debuted on the plains, prairies, and woodlands of the West, poultry keepers have struggled to protect them. From falcons to foxes, weasels to wolves, there has never been a shortage of crafty critters looking for an easy meal.

The chicken, unfortunately, occupies a rather low spot on the food chain. But while our feathered friends might have been short-changed on natural defenses, they do have some deeply ingrained survival instincts. Modern chickens are descendants of flashy little birds from Asia called Wild Jungle Fowl, and they still retain many traits and behaviours that helped their wild ancestors survive and thrive.

Jungle Fowl were especially vulnerable at night and slept high up in trees to evade ground predators. They would seek a roosting site when daylight faded to ensure they were safely settled before dark. Being less-than-proficient flyers, they ascended gradually, flapping and hopping from branch to branch until satisfied with their location. Chicken keepers will recognize this evening ritual—the birds shuffling, jostling, and sorting themselves to find a perfect spot to rest for the night.

Chickens are also hardwired for the "safety-in-numbers" strategy. The more members in the group watching for predators, the better their chances of escaping an attack. There is no mistaking the warning signal from a rooster on guard duty, and when he issues an alarm call, every flock member is instantly on high alert. They scatter for cover or freeze in place, hoping to go undetected until the danger has passed.

But, despite their best efforts to protect themselves, chickens have always been vulnerable, and predators were a round-the-clock concern for pioneers. Farm dogs were important allies in flock protection. Their scent alone was an effective deterrent, and their intelligence and sharp senses made them valuable partners in defending livestock and poultry from attack.

When Mrs. Cookson's Goose was Cooked

Mrs. Cookson raised geese. She was likely aware of the risks of allowing her flock to forage in a nearby field, but her main concerns were probably coyotes and foxes, not a shot-gun-wielding eight-year-old!

John Henry Wood was the son of American immigrants who came to Canada at the turn of the century. Now, in pioneer days, youth were often taught how to shoot at an early age and were quite handy with a gun. So when young John spotted a flock of geese in a nearby field, the prospect of providing meat for his family was too exciting to ignore.

John dashed to the house and grabbed the 10-gauge double-barrel shotgun, but one shell was all he could find. Undaunted, he fired a shot into the flock, but away they flew into the next field. He needed more ammo! John ran to the nearby store to get more shotgun shells for a second attempt at the geese.

This time, he made a stealthy approach, crawling through the field until he was close enough to his target to be sure of success. He braced his gun on a fencepost and let fire. His shots hit the mark, and three big geese went down. Struggling to carry them home, John found them too heavy and realized he needed help. He headed back to the house to fetch his mother.

His mother was overjoyed and filled with pride for her little hunter. And that is when she noticed the geese had leg bands. John had shot Mrs. Cookson's geese! A full confession followed, and John's dad paid Mrs. Cookson for her geese.

Hungry Hawks

Avery Kenny was raised in Toronto, but the call of the West was in his blood. When he met a young, fun-loving seamstress named Alice in 1912, he knew the time was right to find a place where he and his future bride could build a life together. Avery withdrew his savings and boarded the train, making his way first to Edson, then setting out on the Edson Trail—two hundred and fifty miles of mud, muskeg, and mosquitoes.

When Avery and Alice parted ways in Toronto, they had no way of knowing that three long years would pass before they would be together again. But their letters kept them connected, and their long-distance relationship stayed strong.

Avery kept his sweetheart updated on his chicken-keeping pursuits. One spring, he was pleased to write that four of his hens were setting; each had fifteen eggs! He must have had a very successful hatching season, as he later reported having a flock of one hundred chickens. But his good luck did not hold...aerial predators discovered this abundant new food source and began to take full advantage of it. Hawks swooped in and carried away 47 of his chickens, nearly half the flock. Avery tried to thwart them, building a chicken run and covering it by fastening poles across the top. This helped, but the run was not impenetrable. One morning, Avery heard a ruckus in the chicken yard. He went out to find a hawk, brazenly sitting inside the chicken run. He 'made short work' of that hawk.

Newcomers to Canada's great, windswept plains or wooded foothills faced a formidable test. Those who met the challenge created new lives for themselves and their families, building homes in little clearings in rugged northern regions or erecting a cluster of buildings in a sea of grass. They all shared a dream of becoming successful farmers or ranchers, and all lived with constant threats, including the hungry creatures stalking their birds and animals.

Flock keepers did their best to shelter and safeguard the fowl that were so vital to their livelihoods. And, despite menacing carnivores, extreme temperatures, and primitive conditions, their wonderfully adaptable, resilient, and versatile breeds flourished.

Chapter 2

The Breeds That Built The West

October, 1903

Winter had asserted herself overnight. The farmyard was transformed from drab gray-brown to a glittering, twinkling landscape of white. Isabelle kneaded bread dough at the kitchen table. She worked the dough absently, rolling and pressing out air bubbles as her eyes took in the wintery scene outside. Her chickens had ventured out of their coop with tentative steps. The thin layer of snow was unfamiliar to them, but they banded together for courage, leaving a jumble of tracks to mark their progress toward the house.

Isabelle never tired of watching her pretty flock. Jonathan had purchased the chickens back in the spring when he and Isabelle first arrived in Lacombe. He'd seen a notice posted at the general store, "Young Buff Orpingtons for sale." He paid a visit to the farmer, and later that day surprised Isabelle with a crate of seven chickens. They were young indeed, but well past the baby chick stage and dressed in a fine set of soft, honey-blonde feathers. The male had the awkward look of an adolescent—lean and leggy with feet that seemed too big for his body. Isabelle had been delighted.

She had mentioned more than once that they would need chickens for their homestead, and this was a good, reliable English breed she was familiar with. The chicks thrived under her care, growing into adults with soft, abundant plumage as bright as polished brass. Isabelle had not expected to be so charmed by the cockerel, but he had quickly won her over with his good-natured attitude and polite manners. She decided he deserved a name and started calling him Prince.

Although he remained respectful towards people, Prince had assumed the role of flock leader and protector. At nearly eight months old, he had the confidence and stature of an adult, and he watched over the females with one eye to the sky for flying predators. His presence was an effective deterrent to hawks or ravens, and he would not hesitate to fight to the death if called upon to do so.

But with no cause for alarm on this crisp October morning, the birds settled into their busy day of foraging for grain and seeds on the snow-dusted ground.

⁂

What does a typical chicken look like? In today's world, most people would describe one of three versions: a stocky broiler; a lean white hen with a pale-pink drooping comb; or a small, orange-brown commercial layer. It is a gloomy but accurate snapshot of the modern poultry industry.

This was certainly not the case in the homesteading era. Hybrid chickens were still decades away from development, and poultry production was very much a small farm endeavor, rich with vibrant colour and wondrous diversity.

Choosing Chickens

This great abundance of breeds, patterns, and colours meant farmers were not constrained; they had choices. There were, however, a handful of breeds that had proven themselves especially well-suited to thrive in harsh, demanding conditions. Practicality reigned supreme, and the chickens of choice were hardy, dependable, and general-purpose. This meant they had been bred and developed for the ideal blend of egg production, meat attributes, and other key features such as vigour, fertility, and mothering instincts.

Alberta's Department of Agriculture acknowledged that personal preferences should be the first consideration when deciding on a breed. "If merely as a recreation, any breed, from the smallest Bantam to the largest Cochin or Brahma, is equally suitable so long as it meets the fancy of the breeder. If the methods advocated elsewhere in this bulletin for selecting breeders for strain building and for the development of the breed are followed, one will be well repaid by the progress made, no matter what the breed."

But keeping fancy fowl for recreational purposes was not the goal of practical farm folks. They needed sturdy, utilitarian flocks to put food on their tables and contribute, at least modestly, to the profitability of their farming operation. Four outstanding chicken breeds came highly recommended for just such purposes: Plymouth Rocks, Wyandottes, Orpingtons, and Rhode Island Reds.

Plymouth Rock

"If there is a better breed for the farmer, or for those who desire both eggs and chickens, we have failed to find it although many have tried and found wanting."

This was a description of Plymouth Rocks, according to the American Standard Poultry Book, published in 1885. High praise for a breed that was still relatively new, and a foreshadowing of the iconic status Rocks would achieve in North America, even helping to lay the foundation for the broiler industry. Although some traits described by the author would later change (standard weights were reduced, and eggshell colour became consistent), the breed's many attributes gave rise to its importance and immense popularity.

"As table fowls, they have no equal in America; being exceedingly sweet, juicy, fine-grained, tender, and delicate. As spring chickens, they are the very best breed, for, added to the excellence of their flesh, they feather early, and mature with remarkable rapidity. As market fowl they are unsurpassed, being large (cocks weigh 9-11 pounds, hens 7-9) and very plump bodied, with full breasts, clean, bright yellow legs, and yellow skin; they always command the highest price.

As egg-producers, they are only excelled by the Leghorn class, and lay more eggs than any other breed that hatches and rears its own young, and can be depended upon for eggs all the year round. Their eggs are also of large size, very rich, and fine-flavored, from white to reddish-brown in color. In hardiness, both as chicks and mature fowls, they are also unequaled, and being out-and-out an American breed, they adapt themselves to all climates and situations better than any other breed.

In fine, this comparatively new breed combines all the sturdy and excellent qualities of the ideal fowl to a

> wonderful degree, filling a place long sought for, but never
> before attained, and is a golden mean."

By the early 1900s, the Plymouth Rock had become a mainstay on Canadian farms, and the barred pattern, in particular, became synonymous with practicality. They were easily recognizable, with thin, sharp stripes of alternating light and dark traversing the curves and contours of their deep, well-muscled bodies. The local Ag Department also recommended Plymouth Rocks in White, Buff, Partridge, Silver Penciled, and Columbian varieties.

Wyandotte

Another American breed was quickly gaining notoriety in poultry circles. The Wyandotte was not yet well established when the American Standard Poultry Book was published, so its description was brief compared to that of the Plymouth Rock. But it spoke to the beautiful form and valued qualities that would make Wyandottes soar to the greatest heights of popularity. One of the breed's great advantages was its compact rose comb which resisted frost damage. The original Wyandotte pattern was Silver Laced (sometimes simply referred to as Silver). The striking contrast of crisp black lacing along the edge of each white feather defined the breed for many, although other, equally enticing colours and patterns soon followed. By 1917, Alberta frontiersmen were encouraged to select from among seven beautiful and popular varieties: White, Buff, Black, Silver, Golden, Silver Penciled, and Columbian.

> "This new breed has so many points to recommend them,
> both to the fancier and farmer, that they will surely become
> very popular. Their plumage is white, heavily laced with
> black, the tail alone being solid black; the lacing on the

breast is peculiarly handsome. They have a small rose comb, close fitting; face and earlobes bright red. Their legs are free from feathers and are of a rich yellow color.

They are extraordinary layers, surprising every breeder at the quantity of eggs they produce. If allowed to sit they make most careful mothers, are content anywhere, and will not attempt to fly over a fence four feet high. Their great beauty and good qualities will make for them a host of friends wherever the breed is introduced."

~ American Standard Poultry Book, 1885

Silver Laced Wyandotte, Artist, F. L. Sewell, Reliable Poultry Journal 1896, Public Domain

STANDARD WYANDOTTE SHAPE—MALE.
"A Composite Head From Live Models"—As Submitted by the Reliable Poultry Journal for the Criticisms of Judges and Breeders.

Buff Orpingtons Artist F.L. Sewell, Reliable Poultry Journal, 1911

Orpington

One breed from the English class was included in the top four recommended breeds for commercial poultry production. Orpingtons had immigrated to the prairies, bringing majesty and grandness in three solid colours: buff, white, and black. Although Blue Orpingtons were gaining a following, they would not be recognized by the American Poultry Association as a standard variety until 1923. Some breeders raised Rose Comb White Orpingtons in the years leading up to the Great War, but the APA recognized only single combs, and in time, the rose comb lines fell out of favour.

Orpingtons were considered very good layers, producing a steady supply of brown eggs for such a large bird, but their true métier became apparent when dressed for the roasting pan. Their heavy frame, strong build, and well-fleshed breast made them prized meat birds. People also admired the breed's loose, luxuriant plumage, hardiness, and calm nature. All told, the Orpington was a winning combination of beauty, utility, and temperament that quickly captured the fancy of farmers, breeders, and exhibitors.

Rhode Island Red

The quintessential Rhode Island Red was the last of the 'big four' utility breeds of the West. 'Reds' were a new arrival, having just made their debut as an APA standardized breed in 1904. But they made an instant and lasting impression on homesteaders with their distinctive body shape, excellent utility traits, and intense colour. They were a dark red sensation, sleek and lustrous with glossy, iridescent tails. Their friendly, docile demeanor matched their fine appearance, and the beautiful, functional Rhode Island Red swept the prairies like wildfire. They quickly established themselves as model farm chickens, then rose to stardom as showy, competitive exhibition birds.

Rhode Island Reds Print, Source Unknown

The Mediterraneans

These four breeds were considered the major players of turn-of-the-century farmsteads, but there was a large, diverse—and essential—supporting cast. The Leghorn deserves special mention as the embodiment of efficiency and style. They provided an abundant egg supply while bringing grace and jaunty spirit to pioneer farms. Each colour variety (Brown, White, Buff, Black, and Silver Duckwing) was available in either a single or a rose comb version, although White and Brown Leghorns remained most prevalent. Whites were all graceful curves, lean and dignified, where Browns had a more conventional farm-chicken appeal. Leghorn hens were athletic, but not high-strung. Plus, their laying cycle was not likely to be interrupted by broody behaviours, so they continued to fill egg baskets while other breeds were taking a break. In the brown egg-dominated North American markets, Leghorns catered to the minority of consumers with a preference for large, white-shelled eggs.

Another admirable egg producer was the Leghorn's larger, more powerfully built relative, the Minorca. Minorcas could be found in gleaming black, elegant buff, classic white, and with both comb types. Though they combined egg prolificacy with respectable market weights, Minorcas did not quite make the cut as ideal homestead chickens. Hens were known to be flighty, and roosters had oversized, pendulous wattles that were pretty much guaranteed to dip into water buckets and freeze during winter. Frostbite was a concern in the rustic frontier environment, as damage to the sensitive comb and wattle tissues could cause infection and reduced fertility. In extreme temperatures, frostbite was difficult to avoid, but some breeds were naturally more resistant than others.

Heavy Breeds

Some producers were less concerned with stellar egg production; only a plump, meaty, roasting bird would win their approval. The Asiatic Brahmas and Cochins fit the bill—and filled the serving platter as the delicious centrepiece of a special meal.

Their appreciation went far beyond food value, with the Brahma being dubbed 'King of all Poultry'. This regal title portrayed both the stature and the historical importance of the majestic fowl. Although the standard weight of an adult male was 12 pounds, Brahma roosters frequently reached astounding sizes, and 17-pound roosters were not unheard of!

Brahmas were not only powerfully built; they were – and still are - splendidly patterned. The simple colour labels of 'Light' and 'Dark' fall woefully short in describing the precision and striking contrasts of the plumage. Light Brahmas sport the Columbian

Light Brahma, Biggle Poultry Book 1895

pattern—a white base with sharp, black neck striping, wing markings, and tail feathers.

In the natural bird world, males are usually bright and showy, while females are left shortchanged. Not so with a Dark Brahma hen. She embodies strength and beauty combined, being lavishly adorned with a triple set of ultra-fine, concentric feather markings.

Poultrymen praised the Brahma as good foragers, reliable layers of large eggs, and diligent mothers. Their calm disposition made them family favourites, but perhaps their number one quality was winter hardiness. They surpassed the Cochin in this regard due to their practical pea comb, which was less susceptible to the effects of cold and exposure. The magnificent Brahma chicken presided over farmyards with pomp and splendor and would remain a favourite of fanciers for decades to come.

The Cochin was another of the beloved, and ancient, Asiatic breeds. Swathed in a soft profusion of feathers from head to toe, they charmed their caregivers with tranquillity and exuded grandness. The hens were also famously dependable brooders and mothers. All combined, the attributes of these gentle giants cemented their place as enduring favourites of the avian world.

The Java, one of America's older breeds, was a dual-purpose, free-range forager with an early presence in the Canadian West. Their solid build and versatility were prized traits and key ingredients in the development of newer breeds. Javas were instrumental in creating the Jersey Giant, Plymouth Rock, and Rhode Island Red—breeds that would ultimately eclipse the Java in popularity and lead to its unfortunate decline.

Other prevalent—and important—breeds in North America did not seem to gain a strong foothold in Alberta, at least not during the 1900 – 1920 timeframe. The absence of Sussex, Buckeyes, Langshans, and Cornish on the prairies was, perhaps, a consequence of regional availability. Nevertheless, poultry exhibition reports revealed an astonishing abundance and diversity of chickens being raised in and around the western provinces: the flamboyantly crested Polish and Houdans; lovely, spritely Hamburgs; Campines, Anconas, Dorkings, and Games. All were plentiful, and they brought their unique character to farmyards and fairs in a marvelous myriad of color!

Big Benefits of Little Bantams

The miniatures of the poultry world were not as utilitarian as their large fowl counterparts, but they came with two distinct advantages: they consumed less feed and needed less coop space. Poultry manuals of the late 1800s recommended bantam chickens as children's pets, and they were pitched as a good fit for town dwellers with small yards.

Bantam chicken breeds were available in a host of colors and patterns, and their small size made them easy to handle and transport. So, for exhibition purposes, these little birds held big appeal. Bantam Cochin hens also had the advantage of staunch

Bantam Chickens, Biggle Poultry Book 1895

broodiness and were peddled as perfect little incubators. Bantam breeds did provide households with the practical benefits of eggs and meat as well, albeit in smaller packages.

Whichever reason a family might have for choosing to raise bantams, these lively, often feisty little creatures were guaranteed to be entertaining.

New Poultry Pending

Throughout the 1920s and '30s, North America's poultry industry would continue to build on the success of the general-purpose farm fowl. Concentrated selection of high-performing Rhode Island Reds would lead to the development of the New Hampshire, an even faster-growing, earlier-maturing utility breed. At a government research centre in Maryland, Dorkings, Rocks, and Leghorns were being blended and concocted into a heavy, economical breed called the Lamona. By 1921, a Trappist monk in Quebec was celebrating the success of his wondrous, winter-hardy creation, the White Chantecler, while on the other side of the country, Dr. Wilkinson of Edmonton diligently crafted his own cold-tolerant breed. Wilkinson would release his 'Albertans' in an exquisite partridge pattern. These advancements and others shaped the next phase of poultry agriculture...but these are stories for another time.

Opting for Turkeys

Of the domestic bird species being farmed around the turn of the century, chickens were the most numerous by far. By 1921, chickens represented over 95 percent of all Canadian poultry, however photos from the era have captured the odd Bronze Turkey mingling among a group of chickens. A few farming operations specialized in the big, sociable birds.

The Biggle Poultry Book was published in 1895 and featured a section on turkey farming. It was part of the Biggle Farm Library—a collection of immensely popular guidebooks that covered everything from growing an orchard to raising cows. The poultry book was a compact, beautifully illustrated little manual on managing farm flocks. The author, Jacob Biggle, introduced the section on turkey farming with some thought-provoking insights. He suggested that raising turkeys was more difficult than chickens, "because of the differences in the nature and habits of the birds. The turkey is not as thoroughly domesticated as the chicken, having been under the controlling influence of man but a comparatively short time and still retaining many of its wild traits. Their love of freedom, their roving habits and their shyness all indicate their recent introduction from the forest to the domesticated state."

Bronze, White Holland, or Black turkeys were promoted as top picks for the Alberta homesteader, with all reputed to be hardy and well-suited to the rigorous frontier setting. Serving turkeys for special occasions was a tradition already entrenched in Western culture, making turkeys profitable and easy to market.

William Allison was another popular 19th-century author. He declared it unnecessary to describe the illustrious tom turkey... and followed that claim with a most vivid portrayal: "To describe the domestic turkey is superfluous; the voice of the male; the changing colours of the skin of the head and neck; his proud strut, the expanded tail and lowered wings, jarring on the ground; his irascibility, which is readily excited by red or scarlet colours, are points with which all who dwell in the country are conversant."

Western farmers found the birds to be a good match for the climate as they could tolerate inclement weather, sometimes to a fault. The flock would become accustomed to roosting in trees or on the peak of the barn roof. This was probably fine for much of the year, as they were relatively safe and comfortable at night.

But as the seasons turned and the weather grew cold, the turkeys would need to be captured and convinced to sleep inside. That was easier said than done!

Turkey hens were just as tenacious when it came to brooding. Sometimes, a hen would be so reluctant to leave her nest that she would suffer from malnourishment. It was common practice to keep water and food within reach of a setting turkey hen to encourage her to stay fit and healthy while incubating her eggs.

The advent of well-balanced commercial feeds was a game-changer for turkey farmers. The nutritional needs of young turkey poults were complex, and meeting those needs was quite a task in the early days. A varied, but protein-rich diet was key. "The little ones for some hours will be in no hurry to eat; but when they do begin, supply them constantly and abundantly with chopped eggs, shreds of meat and fat, curd, boiled rice mixed with lettuce, and the green of onions. Melted mutton suet poured over barley or Indian-meal dough, and cut up when cold, is an excellent thing. Little turkeys do not like their food to be minced much smaller

Woman Feeding Turkeys, Peace River, Alberta

than they can swallow it; indolently preferring to make a meal at three or four mouthfuls to troubling themselves with the incessant pecking and scratching in which chickens delight. But at any rate, the quantity consumed costs little; the attention to supply it is everything."

The Turkey-less Supper

A long-standing North American tradition is the community turkey supper. The emphasis of such gatherings was to bring folks together and foster a sense of belonging and connection. There was usually a fund-raising aspect, where proceeds might go towards building a church or school. Regardless, the meal was always affordable and provided a much-needed social outing for community members.

In a 1976 interview, Mrs. Mary Edey spoke fondly of such gatherings. She had grown up surrounded by siblings and cousins, so it is no wonder she found the solitary pioneer lifestyle very lonely. In 1914, Mary and her husband were living in an isolated region of Alberta, far from any town. They did not yet own a buggy, so travel was difficult. But they always found a way to attend the annual fall suppers to celebrate the harvest season with other families in the district.

The local Ladies' Aid group organized the first of these events in the area. A few of the members who lived closest to the town held a meeting to make the arrangements. They set the date, organized a menu, and spread the word, requesting that local farmers supply the turkeys for the meal. The plans were all in place; however, a problem surfaced at the last minute...no one in the region raised turkeys.

Mary described how she and other neighbouring farm women came to the rescue with a quick switch of the main course from turkey to chicken because, luckily, most people raised chickens.

"So, you got them ready and you dressed them and you stuffed them and you made pie. And people would come from miles around. Miles and miles. And the old ladies were really good cooks, the older ladies. Oh, when I came out here first I just marvelled at the cooks because I hadn't cooked like that. Plus, it was the pleasure of having company."

Worthy Waterfowl

Settlers coming to this untamed land were welcomed with an extraordinary abundance of wild waterfowl. Canada Geese filled grey April skies with distinctive V-shapes, announcing their return in a rousing chorus. Ducks occupied every span of lakeshore or marshy slough. They dabbled and dived, dipping beneath the surface in search of aquatic plants or snails, leaving only a wedge of tail exposed.

As farm stock, domestic geese and ducks were far less prevalent than chickens, but they represented another viable branch of agriculture. In 1915, the Alberta Poultry Association addressed duck-farming with this statement: "If you have ever tried this branch of poultry raising and failed, it will be safe to say that it would be difficult to convince you that ducks are easy to raise. Perhaps it is also safe to say that not more than one person in every thousand who raises poultry thoroughly understands the care of ducks. Few stay with the business long enough to learn. And yet they are easy to raise when one knows how. In fact, they are much easier to raise than either chickens or turkeys. They grow more rapidly and when properly cared for are free from disease, which I consider very much in their favour."

Perhaps these words, although only moderately encouraging, hit home. By 1918, the Alberta government reported that duck farming had become a rapidly growing industry in the province. For meat purposes, they recommended raising the heavy-weights:

Pekin, Aylesbury, Rouen, or Muscovy. Cayuga and Crested Ducks were not quite as weighty but were still classed as general-purpose market fowl. Indian Runners were much lighter and valued for egg production. One authority declared the Aylesbury the most valuable of the English breeds and the Cayuga the finest of the American breeds. Jet black Cayugas were expected to weigh between eight and ten pounds, were good layers, and were considered comparably easy to raise.

Although ducks required only basic housing and fixtures, not every farm (or farmer) was well-suited for raising the amphibious fowl. The American Standard Poultry Book stated, "It is not in all situations that ducks can be kept with advantage; they require water much more, even, than the goose; they are no grazers, yet they are hearty feeders. Their appetite is not at all fastidious; in fact they eat most everything, and eat all they can. Ducks are the best save-waste; even the refuse of potatoes, or any other vegetables will satisfy a duck, which it thankfully accepts, and with a degree of good virtue. Confinement will not do for them; they must have room, and plenty of it, also a large pond or stream. If you have these requirements they can be kept at little expense."

Free access to a water source was considered fundamental in raising healthy ducks, allowing them to preen and bathe at will. This was well and good for property owners fortunate enough to have a small stream trickling through their home quarter, but supplying fresh water daily was not practicable on every farmstead.

Geese were, in some environments, the more suitable waterfowl. They also needed plenty of fresh water, but could get by without a natural pond or stream. There were, however, other potential issues in having the heavy birds roaming the property. Geese were grazers; they thrived in free-range conditions, and a flock of geese could be very destructive to gardens, farm crops, and even young trees.

Mrs. William Hawthorne feeding chickens and geese, Viking area, Alberta.

One farm manual recommended keeping geese contained within a wooden livestock corral by hanging a "yoke" across their breast. This would prevent them from leaving the enclosure through the gaps in the wooden fencing. The writer of the manual did not recommend keeping geese confined with turkeys or chickens, as a goose could become "very pugnacious" and harass the other poultry!

Three heavy breeds ranked highest as viable commercial options: Toulouse, Embden, and African. The trick to producing a good number of goslings in a single season was to put turkey hens on incubation duty. A well-fed goose would continue to lay if her eggs were removed, and turkeys were expert hatchers.

The massive size of a mature goose was a great advantage in fending off attacks, at least from smaller predators. Hawks or skunks would be intimidated by a goose or gander's hissing, honking, wing-flapping show of aggression and go in search of easier prey. Still, a coyote or fox was a formidable opponent, so poultry guides advised confining the flock at night. Shelter from wind, rain, and cold temperatures was a must, especially for younger, immature goslings.

Profits in Purebreds

Not everyone kept purebred flocks. In fact, the most common fowl on early pioneer homesteads were crosses—the result of hens and roosters of various breeds, mixing and mingling and mating at will. Mixed-breed chickens were cheap to buy and readily available. They were bright and lively, a kaleidoscope of colours and patterns that made a welcome contribution of eggs and meat to the household.

It was the standardized breeds, however, that held the promise of profitability. They had been developed over time through purposeful and intentional mating decisions. Purebred stock offered predictability. Breed standards were in place to ensure uniformity, adding a level of confidence that a purchase would meet the buyer's expectations. Their uniform physical appearance made the breed recognizable, allowing other, less perceptible characteristics to be assumed. Pure stock was, therefore, valued much higher than the common "mongrels" or "scrub" chickens.

A.W. Foley, Poultry Superintendent for the Alberta Department of Agriculture, championed purebreds at every level of poultry production.

> "It is scarcely necessary to state that pure bred poultry of any variety, and particularly our commercial breeds, are the most profitable to keep. The tendency to revert is sufficiently strong in the pure breeds and in scrubs this tendency is so pronounced that it is almost impossible to breed successfully for the market type or for egg production. At no age are scrub chickens as saleable as the pure breeds. For meeting the demands of the higher class local markets or for export, scrub chickens are not satisfactory. To breed pure is to mate birds of the same breed, and to mate crossbreeds means the production of scrubs, and

to attempt successful poultry raising with scrubs results in a decided failure."

Raising quality poultry was not a priority for every family, and mixed-breed chickens would always have their place on homesteads and in backyards. Those who aspired to more serious poultry production had a few decisions to make. Would they focus on chickens, turkeys, geese, or ducks? And then, which breed would best suit their objectives? A limiting factor might be local availability, but once seed stock had been located and purchased, the next step was to multiply them. Replacement pullets were needed to maintain even a small flock, and the more serious, entrepreneurial poultrymen and women had bigger goals. Maintaining flocks of any size, and for any purpose, depended on successfully hatching chicks and providing an environment where they could thrive and reach their full potential.

Chapter 3

Hatching, Brooding, Rearing

May, 1904

The scent of rich, fertile soil filled the air. Isabelle knelt in the loamy garden plot; her nimble fingers plucked weeds from between imperfect rows of seedlings. Weeding was a pleasant chore, and freeing her tender young vegetable plants from competition was satisfying. Sturdy pea plants were shooting upwards, but the beans were just beginning to sprout, emerging as bright green stitches in the dark fabric of the soil. Already she longed for the sweet flavour of fresh-picked peas, green beans, and ripe corn. Last year there hadn't been time to prepare and plant a large garden, but turning sod, harrowing, and sowing vegetable seeds had been a priority this spring.

Their garden was in a flat, natural clearing, a short walk west of the house. The barn, horse corral, and chicken coop were to the east. The chickens had never roamed past the house, and Isabelle hoped they wouldn't discover the warm, loose dirt and tasty garden buffet.

The thought of her chickens prompted Isabelle to brush the dirt from her skirt and set off for the barnyard. Three weeks ago, one of the hens had gone missing from the flock. Later that day, Isabelle found her in a dark corner of the barn, crouched low and silent in her nest. As Isabelle reached toward the hen, she was rewarded with an aggressive display of raised hackles and a whining, growling, warning sound. If any doubt remained, a sharp peck to Isabelle's hand made the chicken's intentions abundantly clear – she wanted to be left alone to brood! Isabelle complied, placing a pan of water and some grain nearby, stopping by once a day to check on her.

Isabelle entered the barn, startled at first to see the nest abandoned and littered with eggshells. But the new mother was nearby, fussing about and keeping her cluster of downy, pale blonde chicks gathered close. The hen spoke to them in soft, gentle clucks, and the babies responded with a chorus of peeping as they mimicked her pecking and scratching motions.

Isabelle's heart filled with joy, and something more...a sense of permanence and belonging. She and Jonathan had made it through their first winter in Canada, and this new generation of Buff Orpingtons signified continuity. Their new life in this land and her lovely flock, all bright and golden as sunlit wheat fields in September.

* * * * *

Hatching Naturally

Chickens appealed to homesteaders because they were inexpensive livestock and their needs were basic. They reproduced with little intervention if left to their own devices, but manufactured incubators had begun to appear in agricultural newspapers and periodicals. They were, without doubt, interesting to learn about, but most stockmen considered such a contraption non-essential for their mixed-farming operation. Why spend money on an expensive piece of equipment when a good mother hen was quite proficient at incubating eggs and brooding chicks?

The broody hen. She is possibly Mother Nature's most marvelous—and most irritating creature. Also known as a "clucky" or "setting" hen, she has been transformed by hormones into a pertinacious bundle of maternal instincts. Her nest might be overfilled with egg contributions from her pen-mates, or it might be empty. It makes no difference to a mild-mannered laying hen who has suddenly morphed into Mighty Guardian of the Nest.

The conduct of a broody hen can sometimes seem annoying, but it is a fundamental component of successful reproduction. Some breeds of hens are more inclined towards motherhood than others. But when the mood to brood hits, the hen will seek a suitable nesting spot. This is usually a sheltered area where she feels protected and secure. Her hideaway might be in a tangle of tall grass under a thorny bush, or she might claim a nest box in the henhouse. Once she has made her choice, she will form her nest with care, lining it with the feathers she plucks from her own breast. She will return to lay an egg nearly every day, but until she has amassed a clutch, the eggs will lie still and dormant. Once she has stockpiled around 8 – 13 eggs, she will decide the time is right to begin incubation.

An egg is truly a wonder of nature. Inside the protective shell is a complete life-support system for the developing embryo. The mother hen provides the perfect environment for the miraculous transformation from embryonic disc to fully developed, ready-to-hatch baby chick. The bare skin of her plucked breast provides the tiny embryo with constant temperature and moisture. The hen is prompted by instinct to turn and arrange her eggs for healthy development. A good broody hen will stay devoted to her task, leaving only for short breaks to eat, drink, and relieve herself. For three weeks, she sits in solitude, content in her sequestration, steadfast to her purpose.

After 20 days of incubation, the baby chicks have outgrown their living quarters. Their safe house has become a prison, and they prepare to escape! The hen can sense the stirring of the chicks, and she stops turning her eggs to allow her babies to orient themselves for hatching. An air cell is located in the blunt end of the egg, and the chick must be properly positioned for its egg tooth to penetrate the membrane, allowing it to take its first breath.

The mother hen can now hear her chicks peeping inside their shells, and she speaks to them in gentle tones to coax and encourage them to free themselves. Hatching is exhausting work, and for the next several hours, the chicks will alternate between resting and chipping away at the shell. Eventually, they will have a jagged, circular rent around the top of the egg, and the hatchlings can kick free, wet, worn out, and wanting only sleep.

The hatchlings will soon be dry and active, but completely dependent on their mother for warmth and protection. She will fluff her feathers to trap heat and make room for them to gather. Her soft clucks and coos comfort them, while quick, excited clacking says, "gather 'round, there is food!" She will break edible foods into tidbits and drop them repeatedly to teach her young which things are good to eat. In the weeks and months to come, she will do her best to teach them skills for survival and keep them from harm.

The mothering instincts of a hen are admirable. Hens will often put themselves at risk to bravely and aggressively defend their young. There are times, however, when those instinctive reactions to danger do not kick in. If a hen does not feel threatened herself, she might not realize she is placing her vulnerable youngsters in a disastrous situation, and disaster was sometimes an unfortunate part of the homestead experience.

A pioneer woman named Pauline was a proficient flock keeper who shared her observations of raising chickens, proving that all experiences, good and bad, are part of the learning process.

Pauline's Precocious Hens

George and Pauline (Hamann) Betts moved from Elkhorn, Nebraska to Canada. Their homesteading experience began in 1903, after two years of planning and preparing. It all started with George and his friend Gust Sachs, two young men with a sense of adventure and a yearning to try something new. Enticing brochures from the Government of Canada spurred them on, and a scouting trip to Canada's Northwest Territories soon followed. They bought train tickets to Siding 14, a community along the C.P.R. line that would later become Ponoka, Alberta.

The viewing was a success, and land purchases were secured. The men returned to the United States to begin preparations for a return journey with their families and belongings. In February of 1903, both families arrived at Ponoka, this time well-equipped with furniture, machinery, and even a pair of mules. But February weather in Alberta can be unpredictable, and bitter-cold tempera- tures greeted them. Pauline and her newborn baby remained in town, but the men were determined to scout out their properties. They realized their mules were not cut out to handle such extremes, so they bought a sturdy team of oxen and hit the trail. The bone- chilling cold did not relent, dropping to a brutal -40 degrees overnight. The men managed to survive the night by burrowing into the oxen's hay stash. They safely reached their properties the next day.

Pauline joined her husband on their homestead as soon as a house had been built, and they quickly settled in. Chicken chores were part of the endless, daily work cycles of pioneer women, but often times it's the simple, everyday activities that become cherished memories. Pauline's narrative speaks of frustration with her vexing mother hens, but her tone is one of tender reminiscence.

"The women too had trying times among their daily tasks. Perhaps the most trying and discouraging part of this was raising chickens, as their sole egg supply and much of their meat depended on how many chickens they could raise each spring. Hens did not lay in the winter so when the weather began to get warmer, the first eggs were eagerly sought. This was not easy as those old hens would make a nest in the most unusual places, such as under the horses' feed boxes, in the hay loft, under some building, in a corner of the pig house, or out in the bush—just about anywhere but in the henhouse. To make matters worse, she would sneak quietly from her nest and start cackling loudly out in the middle of the yard so no one knew where she had laid her egg. It was usually the children's job to search for eggs, but if a hen was lucky and her nest escaped the notice of the children, coyotes, magpies, skunks and the pig, she would lay a nest full of eggs and then set on them for twenty-one days. Shortly after this she would appear in the farmyard with a brood of downy chicks following her. Often the family would catch the hen and put her in a little house with just enough room under the door for the chicks to run around outside. The old hen would be kept there until the chicks grew stronger. If left on her own, the old hen often went to the pig pen, where there was plenty of food, and there, more often than not, while the old hen was squawking and flying around, the old sow calmly ate the chicks, one by one.

[Image 301: "Mrs. Roy Benson (Verna) and children with chicks, Benson homestead, Munson, Alberta.", 1912, (CU1128768) by Unknown. Courtesy of Glenbow Library and Archives Collection, Libraries and Cultural Resources Digital Collections, University of Calgary.]

"Tom Roycroft feeding poultry, Shanks Lake, Alberta.", [ca. 1915], (CU1117997) by Unknown. Courtesy of Glenbow Library and Archives Collection, Libraries and Cultural Resources Digital Collections, University of Calgary.

The more docile hens would lay their eggs in the henhouse, where they were gathered each day and carefully kept from getting chilled. When a hen would start clucking and stay on the nest all day, she would be moved to a nest in a quiet corner and given about a dozen of these eggs. If she was a good hen she would settle the eggs in a neat circle under her and keep them covered for the required three weeks, getting off only for feed and water and being careful not to let the eggs get chilled. There were other kinds of hens too. Some refused to set anywhere but in their own nest, others would set in their corner for a week or so and then desert the eggs for good. Of course, this ruined the eggs and they had to be thrown away. These are only a few of the obstacles that had to be overcome before that lovely fried chicken appeared on the dinner table or the egg on the breakfast plate."

Broody Blues

A good, reliable, broody hen was worth her weight in gold to a pioneer chicken keeper—when her incubation services were needed. Sometimes chicks were not the priority—eggs were—and while a hen was setting, she was not laying eggs. Instead, she would spend her day obstructing nests and hoarding eggs, quickly becoming a non-productive nuisance. There is also a physical toll on hens that brood continuously. Fortunately, poultry keepers had a few tricks up their sleeves.

A 'swinging coop' was often effective in changing the mindset of a clucky hen. It was a simple box, two or three feet square, made entirely of wooden slats. This compact coop would be outfitted with food and water containers and suspended from the ceiling of the scratching shed. Hanging a foot or so off the floor prevented the hen from feeling settled. After a couple of days spent swaying in her bare little apartment, the discouraged hen would abandon notions of motherhood and begin to transition back to the business of egg-laying.

Innovations in Incubation

At the turn of the century, artificial incubators were not exactly new technology; they had been under development for some 50 years in North America. The first machines, however, did not gain much traction with consumers. Most lacked precision in regulating temperature and humidity, and all came with a steep learning curve. One manufacturer delivered their unit in a jumble of pieces with complex assembly instructions. The low hatch rates of these early models discouraged users, causing some to abandon the idea of man-made incubators for good.

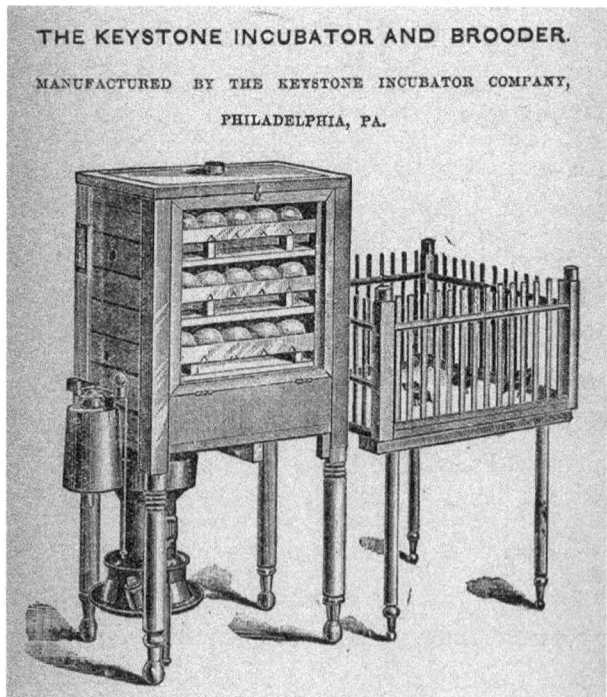

THE KEYSTONE INCUBATOR AND BROODER.

MANUFACTURED BY THE KEYSTONE INCUBATOR COMPANY, PHILADELPHIA, PA.

Keystone Incubator and Brooder, American Standard Poultry Book, 1885

Despite the early mixed reviews, poultrymen of North America recognized the great potential of artificial incubation. An 1885 American poultry manual summed it up nicely:

> "The great question of the day with people going into the poultry business is about incubators. You will find the poultry papers full of those questions and the answers are sometimes quite amusing. One distinguished 'poultry fancier' has never tried them; another, (not quite distinguished) has failed to hatch more than twenty-five per cent of the perfect eggs put in, and still another, a fortunate man truly, hatched ninety-eight per cent of eggs perfect and imperfect. Truly the days of sitting hens are numbered.
>
> But for all these facts in favor of or against the Incubator, it is a settled thing that artificial incubation, attended to by a proper person, will and does pay. Inventors have been taxing their brains for many years to produce good machines, and after many failures, this year of 1884, finds at least a half a dozen good ones in the field competing for popular choice."

As new, improved models continued to be released, operators reported more consistent and reliable results. By 1905, several manufacturers were advertising their wonderous inventions in Canadian poultry journals and newspapers. Government research stations ran tests on different brands and styles to compare their effectiveness. There was a buzz of excitement around these contraptions, and a few adventurous poultry keepers placed orders.

Prairie farmer, H.H. MacPhie must have fallen into this category. His letter to the Farmer's Advocate in May of 1907 describes a typical farm flock of about 50 hens, and although the

primary topic of concern was egg fertility, the writer mentions
using an incubator.

LETTER TO THE EDITOR, FARMER'S ADVOCATE
May, 1907:

One of the best departments of your paper is the poultry
section, where a person can get valuable pointers in the
management and raising of poultry. I keep about fifty
hens during the winter months, and they have been laying
from the first of December up till the present time. In the
morning I feed a warm mash, and about ten o'clock throw
a quart of oats in their litter of straw and chaff, so as to
keep them busy, and about one o'clock or so they get a
good feed of oats, and the third meal comes a short time
before they go to roost, consisting of oats also. A mixed
ration of grain would be better, but having a quantity of
oats in hand, I didn't buy any other feed. My hens are kept
well supplied with green feed, such as cabbages, turnips,
etc., and fresh water is kept before them, and grit, in the
shape of coarse gravel, while sifted ashes serve as a dust
bath, situated in one corner, boarded off. My henhouse has
a large window on the south, which I open every warm,
sunny day; the nests are raised about four feet, and only
one hen can enter at a time. In the matter of hatching, the
last year I used an incubator for early hatching, but did not
have very good luck, the eggs being very poorly fertilized,
but by hen hatching I was very successful, the season being
later.

This season I tried moisture in the incubator up to the time
of hatching, and washed the machine before setting the
eggs in with 10-per-cent solution of Jay's fluid; but did not

have a very good hatch at all, the fault being with the eggs, they being very poorly fertilized, due to my hens being shut in, and a very cold, late spring. I keep about twenty hens with each male. Do you think this is rather too many?

H.H. MacPhie.

Editor's Response:

A cock in hearty condition, with plenty of exercise on a large range, will usually take care of 15 – 30 hens, but in confinement during winter and early spring, half this number be insufficiently attended to. In the case related above it is likely that two males, or else the reduction of the number of females, would have given better results. No matter how many male birds there are, however, results in fertility are not likely to be so good in the early hatches.

- Editor, May 20, 1907 – *Farmer's Advocate and Home Journal,* Winnipeg

Brooding Babies

Chicks needed a warm, protective environment to survive their first weeks of life. The fluffy underside of a mother hen was the ultimate chick-warming device, but with a limited capacity. A large breed hen could shelter up to about a dozen chicks effectively, but anyone looking to ramp up production would need to consider artificial brooding options. Before electricity became widely available, artificial chick brooding was an adventure in itself.

By 1906, the Reliable Company of Quincy, Illinois, had improved on past designs to develop an incubator and brooder combination that ran on kerosene. The lamp was constructed of copper, so it was less likely to break or explode.

Other systems used wood-fired stoves to generate heat. The overhead tank method supplied warmth from a chamber of hot water suspended above the chicks. Longhouse chick brooding employed a greenhouse-type heating apparatus with hot water pipes running through a series of connected brooders. The kerosene lamp brooder was used extensively across the prairies, despite its drawbacks. One negative feature was the need for continuous management; the operator had to keep an eye on the temperature, fill lamps with fuel, and trim the wicks to avoid excessive smoke that could be harmful to the chicks.

Lamp brooders also posed a serious fire risk. Alberta's provincial poultry branch warned users that these contraptions "caused more fires and roasted more chicks than their originators would care to have put on record," so these portable brooders were recommended for outdoor use only. This created another set of issues, especially in inclement weather. A. G. Gilbert managed an experimental farm in Ontario but wrote columns for newspapers across the country. His articles were entertaining and informative, and his exploits with a "Wooden Mother" brooder made for some fun reading.

THE WOODEN MOTHER

By A. G. Gilbert, Manager Poultry Dept., Experimental Farm, Ottawa.

The Brooder is intended to brood the newly-hatched chickens, and in the great majority of cases it is successful in so doing. But there is also much in the ordinary pattern brooder for the operator to brood over. It is really an ugly quantity to look at and operate. For instance, after heavy showers of rain, or days of wet weather, which frequently occur during the chicken season, there may be one or two inches of water on the grass. The brooders I refer to lie flat on the ground, and to reach the lamp, which must be regularly attended to, the operator has to lie on his side.

Some few years ago I had a lively experience of what I am writing about. It was on Dominion Day afternoon, when a violent thunderstorm began, and was followed by others in quick succession, until nine o'clock in the evening. The rainfall was phenomenal. It was as dark as only a dark night in July can be. I had four or five brooders full of chicks on a grass lawn adjoining the poultry buildings. The chicks were in peril, and it was imperative that they should be looked after. When I came to the grass field on which the brooders lay, there was fully two inches of water on the ground. Down I went, first on my knees, but I could not reach the lamp that way. So over I had to go on my side and then I was able to reach the lamp, which the rainwater would certainly have reached had the brooders not been very slightly raised on blocks. Wet, were you? You bet! However, the chicks were all right. After that I had the brooders placed on legs. Just fancy a woman—and many poultry-keepers are—having such a comfortable experience!

But a panacea for all the doleful conditions and experiences enumerated is at hand. Electricity is doubtless the coming factor in artificial hatching and rearing. Already the apparatus is perfected. It is only a matter of a few days when the wizard agency will be attached to one of our incubators and several brooders. No more kneeling or lying prone in two inches of water or mud. No more

nervous dread of the lamp going out or being blown out.
No more smoky lamps; no more unwholesome fumes. You
touch or turn a button, and the requisite temperature is
quickly secured and kept. Welcome panacea!"

~ May 20, 1907 – Farmer's Advocate and Home Journal,
Winnipeg

Fireless Brooding

Lamp brooders were not always reliable or convenient, but baby
chicks were delicate and needed a warm, stable environment. Some
producers experimented with the "fireless" brooding approach, a
method that relied on natural heat retention instead of lamps or
other heating appliances.

The brooder box was designed with a fitted lid and layers
of insulating material to retain the body heat generated by the
chicks. Its effectiveness was improved if located inside a barn or
chicken coop, where residual warmth from the larger birds or
animals helped maintain the brooder temperature.

The system was most effective once chicks were past the
most tender, newly hatched stage, but even then, success was not
guaranteed. Francis T. Shipman of Birch Hills, Saskatchewan,
was an enthusiastic user of the system, though. In 1907, Francis
wrote to the Farm and Ranch Review to share his experiences
with fireless brooding. He explained that the specially designed
crates were positioned under south-facing windows inside the
brooder house. A diagram and construction details were included
for readers wanting to implement the system. The Shipmans had
found a technique that worked well for them, and their convincing
testimony just might have prompted others to give it a try.

"Those who have never seen chickens raised without either artificial heat or that of the mother hen, would scarcely believe that not only can they be raised, but with fewer losses and, I believe, with better constitutions. In the heated brooder the chicks will be too warm at times, again they will not be quite warm enough, consequently, will trample the weaker ones, sometimes killing large numbers, while there are nearly always a greater number put together than ought to be. The nearer we can imitate nature the more successful we are in caring for fowl.

We must consider our climate also, that the sun is very strong early in March and if a brooding house with south windows is provided, the chickens will grow very fast in the warm sunshine. When planning a brooding house one needs to take into consideration to what extent they are going to go into chicken raising, how many incubators are going to be operated and so make sure to provide ample room.

In early spring, when it is cold, and there is not much green food or insects to be found and the price of eggs is high, I consider it more profitable to take the hen-hatched chicks to the brooding house and return the hen to the laying pen.

Hence, one needs ample room as when one has a large number of winter layers, there will likely be a large number of early setters, and if handled in this way the laying pens do not become so depleted.

The brooding house must be built also to suit the season. If no brooding is going to be attempted until April, it need not be built very warm, one ply boards with paper and a dry roof is all that is necessary. Have a south window for about every fifty chickens you contemplate brooding. Have the house built on skids projecting so that a team of horses can be hitched to haul it to clean ground when necessary. It is also a good insurance policy to have room at the end of this building with stove pipes to reach the other end of the brooding house to take the chill off in case of rainy weather when the birds are deprived of sunshine. I have made tests of this system of fireless brooding versus the mother hen. I can see no difference in the growth of the chicks which I took from the incubator and gave twenty to hens and divided the other in lots of twenty for the fireless brooding test. There are no losses through sickness or accident if penned and fed with reasonable care. Never forget to clean coops and use clean, dry litter. There is nothing worse than a foul coop."

Turkey-Raising Tips and Duckling Details

Those who specialized in raising turkeys or waterfowl expanded their flocks following the same basic principles as chicken farmers. The work of incubation and brooding could be handled by a nurturing mother or provided artificially. Turkey hens were reputed to be very capable and committed parents, but they tended to wander away from the safety of the farmyard. It was, therefore, advisable to keep the hen confined to a small shelter that still allowed her young poults some freedom to come and go.

A-Shaped Brooder Coop for Turkeys

"Many different plans are advocated for raising the young birds, more particularly for the first five weeks. The A-shaped coop with slatted front and without a bottom or floor can be easily moved the breadth of itself on to fresh ground each day; the young will take in as much range around it as is good for them, and it will not be necessary to hunt for the turkeys if a sudden rainstorm looms up, because they will go in the coop of their own accord if it rains hard enough to injure them; and finally, it renders unnecessary the driving into a building at night, all that is required being a broad board to prop across the front of the slats. It is highly desirable to keep the coops with turkeys some distance from broods of chickens."

~W.J. Bell, May 20, 1907 – Farmer's Advocate and Home Journal, Winnipeg

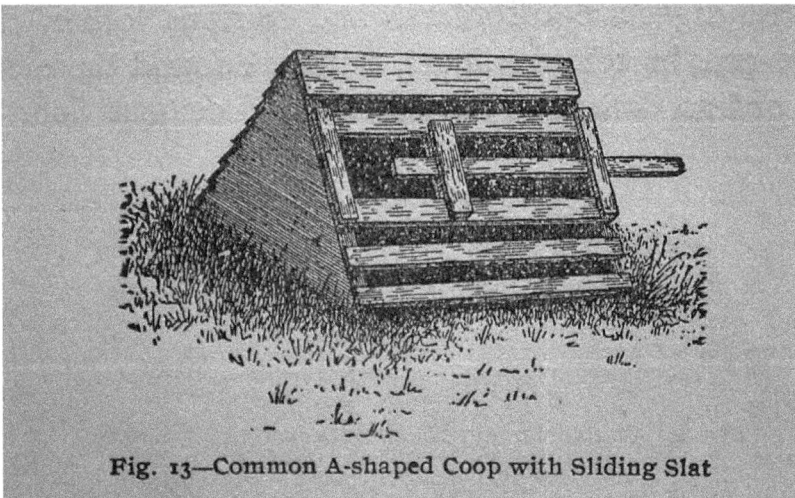

Fig. 13—Common A-shaped Coop with Sliding Slat

The Farmer's Short Courses in Livestock, The F.B. Dickerson Company, Lincoln, Neb. 1921

The Alberta Provincial Poultry Association was founded in 1913 with a broad mission: to advance all types of poultry production. Ducks were no exception, and the club published the following pointers to help people find success in raising the quirky little quackers.

"To begin at the beginning, it is necessary to have fresh fertile eggs to start with. Duck eggs will not endure much rough handling. If they are shipped during warm weather there is not so good a chance to secure a good hatch, besides when a duck egg is ten days old it has reached the age limit, so far as hatching is concerned.

The eggs are thin shelled and therefore easy to test, we test all our eggs no matter how they are being hatched. Duck eggs can be hatched in incubators just as well as by the ducks themselves or by chicken hens. When the eggs are tested there is no excuse for allowing a hen to set on a lot of infertile eggs for several weeks. When these are removed it gives the others a better chance. Be careful in doing the work, as it is very difficult for a beginner to tell a certain stage of incubation, a perfectly good duck egg presents a rather queer appearance to the novice. When held before a strong light the shell appears to be nearly half empty. Don't get excited and throw the egg away, if you do the chances are that you will destroy a duckling."

Once hatched, baby ducks provided no end of entertainment. J. Biggle described them as "animated balls of down, seldom quiet and never so happy as when eating or dabbling in water." Biggle explained that newly hatched ducklings had more strength and vitality than chicks and did not require the same continuous, hovering warmth from their mother. In good weather, he claimed that a group of up to 40 ducklings could be cared for by a single hen duck!

Resources on artificial duck brooding echoed Biggle's advice, with less emphasis on steady warmth. Conventional practice was to confine young ducklings to the brooder at night and return them at regular intervals throughout the day. Moving ducklings in and out of a brooder all day is impractical, but baby ducks do not stay small for long. They double in size in no time, and would soon graduate to larger accommodations.

Learning Through Literature

During this westward expansion period, most settlers came from agricultural backgrounds. These folks were already familiar with farming techniques and knew at least the basics of caring for poultry and livestock. Not everyone was a seasoned farmer, though. People from every walk of life were lured to the prairies by the promise of land and the prospect of building a better future for themselves. Books and manuals could fill some of those knowledge gaps. English was the dominant language of the West, which created some barriers to learning for immigrants with only basic English literacy skills. Most of the books and publications available at local stores were printed in English, including these two popular titles: *Lippincott's Farm Manual* and *The Farmer's Short Courses in Live Stock.*

The Short Courses in Live Stock book was a massive, 959-page compilation of material from the "most successful veterinarians in the world, and the most practical farmers and stock owners in the United States and Canada." The relatively small section devoted to poultry farming was practical indeed, as summarized in these 15 'Poultry Pointers'.

Poultry Pointers from The Practical Stock Doctor, copyrighted 1904 – 1920.

In the raising of poultry or stock of any kind it should be the aim of everyone to keep it healthy and improve it. You can do it very easily by adopting some systematic rules. These may be summed up as follows:

1. Construct your house good and warm, so as to avoid damp floors and afford a flood of sunshine. Sunshine is better than medicine.

2. Provide a dusting and scratching place where you can bury wheat and corn, and thus induce the fowls to take needful exercise.

3. Provide yourself with some good healthy chickens, never to be over three or four years old, giving one cock to every twelve hens.

4. Give plenty of fresh air at all times, especially in summer.

5. Give plenty of water daily, and never allow the fowls to go thirsty.

6. Feed them systematically two or three times a day. Scatter the food so they can't eat it too fast or without proper exercise. Do not feed more than they will eat up clean, or they will get tired of that kind of feed.

7. Above all things, keep the house clean and well-ventilated.

8. Do not crowd too many into one house; if you do, look out for disease.

9. Use carbolic powder occasionally in the dusting bins to destroy lice.

10. Wash your roosts and bottom of laying nests, and whitewash once a week in summer and once a month in winter.

11. Let the old and young have as large a range as possible, the larger the better.

12. Don't breed too many kinds of fowls at the same time, unless you are going into the business. Three or four kinds will more than keep your hands full.

13. Introduce new blood into your stock every year or so, by either buying a cockerel or a setting of eggs from some reliable breeder.

14. In buying birds or cages, go to some reliable breeder who has his reputation at stake. You may have to pay a little more for birds, but you can depend on what you get. Culls are not cheap at any price.

15. Save the best birds for next year's breeding and send the others to market. In shipping fancy poultry to market, send it dressed.

Agricultural textbooks were often displayed prominently on shelves in rustic homes, a trusted and dependable resource with answers to all the most common questions. Poultry reference books were often illustrated with beautiful artwork, inspiring readers and becoming treasured possessions. The less permanent sources of reading material were, however, just as essential to frontier life.

The weekly newspaper was the medium for marketing goods and helping rural residents feel more connected with their communities.

It is not an exaggeration to say that agricultural newspapers were indispensable to the pioneer's existence. This was the main source of industry news—and amusement! Advertisements and event announcements helped those living in isolated areas feel less removed from society. Some papers featured fictional stories that spanned several issues, leaving readers waiting impatiently for the next installment. From sewing projects and children's columns to editorials and building plans, a newspaper subscription enhanced the lives of every household member. In 1911, an annual subscription to a weekly paper such as The Farmer's Almanac and Home Journal cost $1.50...not an unreasonable investment for a year's worth of fresh reading material. Knowing the latest issue was due to arrive at the post office might even have been incentive enough to hitch up the rig and make a trip to town.

Agricultural papers were a trusted guide for aspiring poultry farmers, with regular columns and expert articles focused on fowl. Subscribers could even have specific concerns addressed by a poultry specialist through a letter to the editor. Farm papers were the networking tool of the century, helping individuals transform their small personal flocks into thriving commercial ventures.

Chapter 4

From Pin Money to Profits

August, 1904

The midday sun beat down without mercy. Jonathan swung the sledgehammer again, and the fence post sank a half inch deeper into the hard-packed earth. Isabelle held the wooden post steady, wincing with each pounding stroke.

Jonathan lowered the hammer and tested the post for give, determining it was firmly set. "Let's take a break for lunch. It's hot as blazes out here," he said, wiping his brow with his forearm.

Isabelle was quick to agree. Her face was flushed from the heat, and wisps of hair had escaped her bonnet, clinging damply to her neck. The new calf corral was taking shape, but several days' work remained.

Leaving their tools, the couple walked towards the house. Chickens scattered as they passed through the flock. Jonathan laughed as one hen made a spectacle of loud, indignant squawking and wing-flapping.

Isabelle's flock had expanded over the past months. Three of the six hens had each raised a clutch of chicks, and her original group of seven now numbered 26. The youngest batch had sprouted their first set of feathers; only the last remnants of soft chick down remained on their heads. A pair of gangly young cockerels from the first hatch began a playful sparring match. With hackles raised, they bowed towards each other sporting golden collars of sharp, narrow feathers. But the game ended as abruptly as it started. Their feathers relaxed with the mutual decision to remain friends, at least for the time being.

Jonathan mused, "We'll have to deal with some of these cockerels in the fall."

"Yes", Isabelle agreed. "I'll be glad to have some fresh chicken to cook, but I didn't expect the hens to raise so many chicks. There is only space in the coop for ten or twelve birds over the winter. It might be best if we sell some chicks."

"Good idea," Jonathan answered. "Why don't we write the newspaper and place an advertisement?"

Small Scale Sustainability

The elderly bachelor down the road has extra potatoes and onions, but no chickens. He gladly trades vegetables for fresh eggs. A woman accepts lard and smoked bacon as payment for two dressed cockerels. The teenage neighbour boy has a knack with horses and offers to spend an afternoon ground-training a colt in exchange for a few chicks. The barter system was alive and well on the frontier, and poultry products were always in demand.

Table eggs could also be sold for cash, and countless farm wives relied on egg money as their sole source of income. While

Ad section from Farm and Ranch Review, March 5, 1914

the typical homestead flock did not generate big profits, putting a price on financial independence is difficult. The ability to earn— and spend—money on her own terms was a rare privilege for many pioneer women.

Chickens were the unsung heroes of pioneering. They put food on the table and spending money in women's pockets. During times of hardship, when every penny counted, the humble homestead flock supported farming operations and sustained families. But such small-scale production, while a valued asset to the individual landholder, did not begin to address the needs of the broader population. There was a vast, untapped market for table eggs, dressed poultry, breeding stock, baby chicks, and hatching eggs, waiting to be developed and supplied.

Entrepreneurial-minded folks, both men and women, saw the unlimited opportunity unfolding in the poultry sector and prepared to capitalize. They acquired better breeding stock, scaled up production, and expanded offerings to become bigger players in the burgeoning industry. Promotion was key to success, and agricultural newspapers became riddled with advertisements.

Newspaper marketing varied from a brief line or two in the classified section to stylish display ads. At a rate of two cents per word, classified ads were the most affordable option, and frugal sellers became masters of brevity, providing just enough detail for an interested party to line up a purchase. Rural mailing addresses consisted of only the seller's name and their closest town—no postal code required!

A more serious producer might feel justified spending more on marketing and promotion. Display ads started around $1.12 per inch. Some featured bold borders and elegant flourishes to capture the reader's attention and create a positive impression. A photo of an award-winning bird added a convincing incentive to purchase from the seller's distinguished line.

Expanding the Egg Industry

By 1913, 1.9 million dozen table eggs were imported to Alberta annually. The entire western region of Canada was a net importer of poultry products, with eggs coming from as far away as China. The Dominion had been heavily invested in developing Western agriculture for years, and the poultry sector was no exception. The province was not yet one year old when Alberta's newly appointed, and very optimistic, Poultry Superintendent stated, "There is no valid reason why Alberta should not produce all the poultry and eggs she needs and have some to spare for British Columbia markets where poultry products always command a very high price."

A steady local supply of table eggs did not exist in 1906 Alberta, partly because the average homestead hen was not selected or bred for production. A provincial survey found that the typical Alberta hen laid a meager 50 eggs yearly! Clearly, the Superintendent had his work cut out to build an egg industry that could meet the immediate demands of local consumers, keep pace with the ever-expanding population, and yield a surplus for neighbouring provinces. The Department of Agriculture hatched a plan to ramp up the quality and production of table eggs in a multi-faceted approach, offering instruction, demonstrations, and breeding stock.

Educational work was a top priority, and instructional bulletins were printed, revised, and reprinted. Articles were published in newspapers, and standards for egg marketing were proposed, though not yet implemented. There was no egg marketing board or regulatory process for quality control, and the collection and handling of farmed eggs left much to be desired. Farmers were urged to up their game, but the egg game was barely into its first inning, and the team players were amateurs. So, information dissemination started from scratch—egg farming 101.

"While there are a few egg producers who take the best care of their product, the average farmer considers the eggs produced on the farm a by-product and makes very little provision for their care, aside from gathering them. A large loss is caused by dirty eggs, the number being enormous. This loss is very largely brought about by not gathering the eggs often enough. In wet weather more dirty eggs are found than at any other time. This is caused by the fact that the hen's feet are often covered with mud or other filth, and in going on the nest to lay she soils the eggs already in the nest.

An insufficient number of nests is often the cause of many of the dirty eggs found. Eggs are laid on the ground and around the hay and straw stacks becoming stained, and are classified as "dirties". Again, when too many eggs are allowed to remain in a nest some are broken and many of the others become smeared with broken yolks. This condition is often brought about by allowing the broody hens to use the same nests with the layers. On a farm where one nest to every four hens is provided and the nests are kept clean and well bedded, it is found that very few dirty eggs are produced.

After gathering the eggs, care should be taken not to put them where they will become heated, or near oil, onions, or other vegetables, as they readily absorb odors.

Although dirty eggs may be perfectly fresh, they invariably sell as "seconds", and when but a few dirty eggs are mixed with an otherwise fresh, clean lot, they materially decrease the price of the clean eggs."

Seasonal Shortages

Eggs were a seasonal crop and highly subject to the law of supply and demand. Hens worked hard in the spring, producing half of their annual output between March and June. They were usually at peak production in April, then gradually declined throughout the weeks and months that followed. Winter eggs were scarce and commanded a premium. In 1917, consumers could buy a dozen eggs at the spring bargain price of 18 cents, but the cost could increase to as much as 60 cents in winter. Alberta Agriculture urged poultrymen to bump up winter egg production to maximize their profits and do their part to address egg shortages. This was easier said than done.

One of the main factors affecting seasonal egg production is the hours of daylight. Nature's methodical planning guides the hen's reproductive cycle, beginning with the lengthening days of springtime that stimulate the hen to begin ovulating. If her eggs are fertilized by a rooster and incubated for 21 days, her chicks will enter a world bursting with new plant life and mild temperatures. Summer is the peak time for growth. The chicks will thrive and mature in warmth and abundance before the onset of autumn's cool nights.

Modern chicken keepers use artificial lighting to control the egg-laying cycles of their hens and boost production, a technique unknown to early pioneers. Before the turn of the century, Dr. Waldorf from Buffalo, New York, was studying the influence of artificial light management on egg production, but those in Western poultry circles remained in the dark. This groundbreaking discovery would later revolutionize the business of egg farming, but in the early 1900s, producers had no cut-and-dried methods to guarantee winter egg supplies. Industry specialists offered three key recommendations: quality feed, good flock management, and targeted breeding strategies.

The prevailing theory of the time was that enhanced living conditions would motivate hens to lay more eggs. During the coldest time of year, when hens were confined indoors, flock keepers were advised to make every effort to provide an environment that mimicked their summer lifestyle as closely as possible. In milder seasons, their foraging activities provided plenty of sunshine and exercise, and this should be replicated with a bright and roomy wintertime scratching shed. A large component of their outdoor foraging diet was comprised of fresh vegetation and animal proteins from eating worms, insects, and grubs. Their winter diet should, therefore, include a variety of green foods like cabbage, sprouted grain, clover, or alfalfa, along with animal proteins in the form of meat scraps, blood meal, or bone meal to supplement their daily rations. Skim milk and buttermilk had excellent nutrient value year-round, and special milk troughs for hens were available for purchase. Of course, plenty of fresh water was always a necessity. In freezing temperatures, responsible flock owners carried buckets of warm water to the coop at least twice a day to prolong access to water.

Producers were also advised to hatch themselves a good supply of replacements early in the year, as fresh young pullets were more inclined to continue laying through the long, dark days of winter than aging hens. And again, poultry farmers were urged to supply quality products.

"Keeping hens for summer egg production is antiquated. According to present advanced methods in poultry raising chickens are hatched in spring, the hens are fattened and killed in June or July and the spring hatched pullets lay throughout the autumn, winter and spring months. In this way the eggs are produced at a time when they command the highest price owing to the limited supply at that season of the year, and the advanced poultryman has no surplus supply of eggs in summer when an over supply would trouble him and when they are the cheapest.

As well as endeavoring to supply the eggs at a time when they are in greatest demand the poultryman must get into touch with the best market. The price depends very much upon the way the eggs are presented for sale.

The cook likes to get eggs of one size and color and the larger the better. It behooves the poultryman therefore to endeavor to satisfy the whim. This is done by having but one breed, having it pure and using only eggs that are large and uniform in shape and colour for incubation.

Do not allow the male birds to run with the flock at any other time than the breeding season as they are not in any way conducive to egg production. The better way is to kill and market them immediately after the breeding season is over."

~ Department of Agriculture Poultry Bulletin No. 3

The recommendation to separate the roosters from the hens raised some concerns. Multiple pens and housing were not the norm in farmyards, and an adequate number of males were needed to ensure spring fertility. It was risky for farmers to slaughter their male breeding stock after hatching season and pin their hopes on raising or purchasing good replacement cockerels. The Ag Department offered a solution.

A government-owned poultry breeding facility was established at Edmonton's Hudson's Bay Fort grounds, just south of the Legislature. The poultry plant was later moved to the University farm. In 1908, the facility began to supply hatching eggs and live chicks for purchase. It also included a rooster lending service. Following the premise that infertile eggs were better for eating, they hoped to convince farmers to keep only hens in their laying flocks, borrowing male birds as needed from the Department's 'breeding plant'.

Prairie farmers saw the potential of shifting more heavily toward egg production, but unstable market demands and price fluctuations were concerns. They were also aware that the inconsistent quality of eggs hitting the market was hurting business. They believed these issues could be mitigated by developing an egg producer's cooperative, or "egg circle." This type of program was finding success in Ontario which provided a framework to start up a similar system in the West. A meeting was organized to introduce the idea to the farming community and discuss the potential benefits for Western producers.

Egg Circles for the Western Farmers – September 5, 1912

In Ontario egg circles, having for members the farmers of a specified district, have met with much success within the last couple of years.

A short time ago a similar movement was started among the farmers of the C.P.R. colonies around Strathmore, Alberta.

An enthusiastic meeting was held at Strathmore and plans were made whereby the C.P.R would buy from the farmers comprising the egg circle, all the eggs they could raise, providing the farmers adhered to certain rules and regulations regarding the culling, keeping and marketing of the eggs.

There is no apparent reason why these circles should not be formed in many more of our western districts.

An egg circle among farmers is only another name for co-operation in eggs among them, and besides interesting them in the poultry industry, and there is money in eggs, it brings the farmers into a sort of union, it gets them a better market for their product, it gives them stimulus to seek to improve their breeds of poultry, and if for no other reason than the foregoing, ought to receive their hearty support.

The hens of a farm have been and are too often today, looked upon as a sort of half-necessary sideline, and in many cases the flock receives scant attention.

If they lay, well and good, but they don't, the farmer thinks he will wait and see what they will do "after a bit".

Of course, on many places the farmer's wife is keenly interested in the welfare of her hens, for on them to some extent depends a part of her pin money, and she takes care to do her level best to make her hens pay.

But the case with many folks is that for want of proper attention the hens have got too old. New blood in the shape of the male birds has not been introduced into the flocks and the hens are useless, except perhaps as broilers, and pretty tough ones some of them make at that.

Then, supposing the farmer or his wife do take an interest in their poultry, and many of them do take a great deal of pains with them, when they take their eggs to the country store, do they get in return what the eggs are worth?

If they will take the amount out in trade, they will get value of say 30c per dozen; if they ask for cash, they will get a couple or three cents less, or in some cases the store keeper refuses to accept eggs except in trade. Is the farmer or the farmer's wife getting a fair show?

This is where the egg circle comes in, and its benefits are at once apparent. A good, reliable firm agrees to take all the eggs, or perhaps so many dozen a week from a certain egg circle. The farmers who form this circle agree to sell all their eggs to the aforesaid buyer, who stipulates that the eggs be fresh, that is, not more than seven days old; that they be uniform in color and size, properly packed so that there is no possibility of breakage, and if these regulations are fulfilled the circle receives sometimes three to five cents and as high as ten cents a dozen more for the eggs than would be received at the local store.

Then the farmer or poultry man has the small, badly shaped, culled eggs for his own use and consumption, and they are just as good eating as the others, only to put them with the uniform eggs would be like putting feed wheat with No. 1 hard, and expect to get the highest market price for the mixure.

The egg circle brings the producer and consumer together. It is certain in nearly all districts to meet with strong opposition from the local store men, but if the men who sell the eggs stick together, as the storekeepers stick together, and put on the market the finished article, they are bound to win out." [Farm and Ranch Review, Sept 5, 1912]

The Western Egg Circle proposal caught the attention of Alberta's Department of Agriculture. The department publicized the idea and, in 1913, took on the responsibility of receiving and grading the eggs. It was not a long-running program, though. It had been effective in Ontario's more densely populated farm regions, but western egg producers were spread too thinly to keep this coordinated approach sustainable.

Lucrative Leghorns

Egg marketing boards remained a distant vision, but individuals began to report healthy profits from selling eggs independently. Mrs. Murphy from Cowley, Alberta, raised White Leghorns, and her letter in the Farm and Ranch Review likely inspired many rural women to reevaluate their flock-keeping goals. Her accounting records proved that there was serious money in eggs.

"I was figuring up my poultry account today, and as I think it a little unusual, I thought I would let all poultry lovers know the possibilities of poultry keeping, and to tell them that success is not so difficult after all. Here are the figures:

Stock on hand Jan. 1st 1913
100 hens 150.00
Feed Purchased during year..................................146.56
                                                      $296.56
Eggs and poultry sold and consumed 400.03
Stock on hand Dec. 31st 1913.................................. 297.00
                                                      $697.03

"Woman feeding White Leghorns on southern Alberta ranch.", [ca. 1907-1908], (CU1115414) by Unknown. Courtesy of Glenbow Library and Archives Collection, Libraries and Cultural Resources Digital Collections, University of Calgary.

"Manir Polet Feeding Chickens" (ca. 1907): Provincial Archives Alberta Photo A7693

This shows a profit of $400.47. I had to buy all the feed, which consisted of the available grains, oats, wheat, barley and bran and shorts. I kept oyster shell and charcoal before them all the time. I had no green feed in winter and did not feed any meat products. They had plenty of coal ashes at all times. These hens were housed in an ordinary house with south windows. I cared for them in my spare time. As well as this being a nice little profit for an amateur, I assure you it has been a healthful and pleasant pastime."

(Mrs.) I.L. Murphy

Fortuitous Feathers

Eggs, meat, and breeding stock were the three pillars of the poultry business, but feathers could provide an added source of income for anyone willing to package them up and ship them to a buyer. Goose down and feathers were valued most highly, as they made mattresses, pillows, and quilts soft, fluffy, and well-insulated. In 1914, goose feathers commanded a premium of 60 cents a pound for white or 50 cents for mixed colours. Duck feathers brought 35 cents a pound. The tail feathers of turkeys were used in making dusters and were worth 40 cents a pound.

Feather finery was extremely popular in the early 1900s, and a boa of curly white feathers was sure to enchant any fashionista. She may not have realized she was wearing ordinary chicken feathers curled with a hot iron! White feathers could also be dyed to make colourful muffs and fans. Featherbone was made from the quills of wing and tail feathers. It added structure to corsets and petticoats.

Milliners bought chicken feathers by the bundle, using them to create extravagant ladies' hats bedecked with arrays of elegant plumes or even whole bird replicas. Bird hats had been all the craze in Europe since the late 1800s, when exotic, wild birds were hunted specifically for that purpose, some species to the point of extinction. Thankfully, the slaughter of wild birds for hat-making fell out of favour and had all but ended by 1900. With chicken feathers offering an ethical alternative, a lady could avoid controversy and still feel festive with a fancy bird perched atop her head.

Feather sellers were offered the best price if their products were properly sorted, so tail, wing, and 'skirt' feathers were usually sold in separate bundles. They were priced according to length, cleanliness, and quality. Regular body feathers from chickens could be packed in sacks. Five chickens would yield around a pound of feathers, and a shipment consisted of either a 200-pound bale or six food sacks. These "common mixed" were worth about 5 cents per pound and mostly ended up in feather beds.

Fragile Freight

A fertile egg is a parcel of possibility. Lying dormant within the shell are the benefits and advantages of a proven bloodline. Homesteaders were beginning to realize the earning potential of high-performance flocks and were looking for stock from reputable strains. As postal delivery and rail shipping services expanded, the sale of hatching eggs became a viable method of transferring quality bloodlines from one producer to another.

Purchasing fertile eggs involved letter writing, sending payment through the postal service, and awaiting delivery. At last the crate would arrive and be pried open with trepidation, hoping beyond hope the eggs had survived whole and intact, safely nestled in soft nests of hay.

Upon delivery, the eggs would need incubation. The few progressive poultry keepers with manufactured incubators would fire up their machine, but the majority of small farms still relied on a trusty setting hen to do the work. The next three weeks would be filled with anticipation and visions of vigorous, well-bred birds teeming with good health and displaying the desired attributes.

Regardless of the incubation method, purchasing fertile eggs was a gamble, and long-distance travel compounded the risks. Many factors influenced hatch rates, starting with the eggs' quality, freshness, and fertility. Even clean, fresh, high-quality eggs would not hatch if they arrived cracked or scrambled inside the shell from rough handling. Each egg contained a tiny cell cluster that could be damaged, with little external evidence to alert the buyer. The handling of the package during shipping was mostly outside the shipper's control, but favourable results were far more likely with well-packaged eggs.

Best practices on the fine art of egg shipping were summarized and printed in the March 10, 1910, issue of the *Farmer's Advocate and Home Journal*.

"It is a well known fact that eggs for hatching sent by post or rail frequently give poor results. The fault lies sometimes with the eggs, but still more frequently with the system of packing adopted. The aim should be to avoid not only broken shells but also to prevent injury to the delicate membrane enclosing the yolk, as an egg may be completely spoiled for hatching without a trace of fracture appearing on the shell. This can be prevented by using a package of moderate size and weight.

Of the many patent egg boxes some of the best are too expensive, others are too small, and a still greater number too fragile. The popular cardboard boxes are objection-able. They undoubtedly save labor in packing, are light in weight, but their initial cost, the number of breakages that occur whenever they are used, and the fact that so few people return them, make these boxes an expensive item for the small poultry keeper.

After a trial of many different kinds of package, nothing has been found to compare with a plain wooden box 11 x 7 ½ x 3 ¼ inches (outside measurements) made of the very lightest boards."

"To pack a dozen eggs, a layer of hay is placed at the bottom of the box. Each egg is first wrapped in a piece of newspaper and then in a strip of soft hay, after which it is placed on end in the box. A box of the dimensions given holds twelve eggs in four rows of three eggs each. It is most important that the eggs should stand on end, and that they should be so tightly packed that they cannot move when the box is roughly handled or shaken. The proper amount of hay to use is easily determined with a little practice. The lid should be tied on, never nailed, and no label is necessary, as the address can be written with indelible pencil on the white wood. The danger of having valuable high-priced eggs broken or interfered with when sent in a box that is tied only, and not nailed, can be overcome by screwing down the lid.

Every vendor of eggs for hatching should be provided with a stamp and a bottle of endorsing ink to stamp every egg sold. By this means, any attempt to substitute inferior eggs on the journey or to claim falsely for the replacing of infertile eggs can be detected.

In order to get best results, all eggs for hatching that have been sent [on] a journey should be unpacked and allowed to rest on their sides for twenty hours before they are placed under the hen."

The cost of hatching eggs varied, but turkey and goose eggs were generally priced higher than chicken eggs. Mrs. B. Walters of Clive, Alberta, was a seller of all three. Her husband was a well-known breeder of Shorthorns who exhibited his cattle far and wide under the name of James L. Walters and Sons. As James gained recognition in the purebred cattle industry, his wife Beryl, a busy mother of eight, capitalized on raising purebred poultry. She raised Buff Orpingtons, Bronze Turkeys, and Toulouse Geese. In March 1914, she offered a setting of 15 Orpington hatching eggs for $2.00, and turkey or goose eggs at 25 cents each.

March was prime time for hatching egg sales, and that same month, Mrs. Constantine of Carstairs, Alberta, ran a compelling advertisement. She promoted her Single Comb White Orpingtons as "healthy, robust, A.1. layers. Pure Kellerstrass hens, mated to a magnificent, absolutely white cockerel, direct from Aldrich Poultry Farm." The sire of this cockerel was reputed to be a Madison Square Garden winner. Fertile eggs from Mrs. Constantine's highly acclaimed flock were $3.00 for 15. For perspective, farmland advertised in the same issue started at $11.00 per acre, or the equivalent of 55 hatching eggs.

Stock Marketing

Transporting fertile eggs was a convenient way for poultrymen and women to add 'new blood' to their flock. It was one way to access superior strains from across the country or even internationally, but it was no substitute for live breeding stock. Buying live birds from a reputable seller fast-tracked the process of starting up or improving a flock by removing the risks of damaged eggs and dismal hatching results.

Like fertile eggs, live bird prices varied widely, depending on the age of the stock, the breed, and the strain. Another factor that influenced pricing was the time of year. Buyers who practiced patience might find their desired birds available at a discount.

> "July is a month of bargains in the poultry world. Every breeder is cutting down on his stock by cutting prices, and when this occurs it is a good time for amateurs to buy. The beginner who has spent a few dollars for eggs can profitably get a breeding pen now for the coming season. The novice can purchase hens and cocks for showing purposes later at a greatly reduced figure. The breeders demand room for the young stock and will make prices that usually move them out hurriedly. To him who needs old stock or the man lately interested we say: Buy stock in July. Yearling and two-year-old breeding birds are the best to use anywhere, and they are cheap now."

Producers who raised quality bloodlines realized higher profits in selling meat and eggs for consumption. Plus, they unlocked a lucrative third revenue stream—supplying other homesteaders with live breeding stock and hatching eggs. The initial start-up cost of purebreds was higher, but it was an investment that increased the profitability and sustainability of their farming or ranching operation.

But while financial gains were the motivation, the outcomes were far-reaching. Poultry producers were the medium for distributing breeds and bloodlines throughout the Western frontier.

The role of the typical poultry producer remained firmly set on propagation and distribution, separate and distinct from that of the poultry breeder. Breed development was driven by 'craft breeders' (fanciers) and agricultural branches. Craft breeders sometimes worked independently, but many were backed by government or university programs. By 1900, sophisticated poultry breeding facilities were operating throughout the Dominion of Canada.

Although farmers did not typically specialize in poultry breeding, they did utilize proven strategies to maintain or enhance the efficacy of their prized flocks. Their mating decisions amplified or reduced specific qualities and characteristics, molding and refining bloodlines and, potentially, entire gene pools. Traits were strengthened, and unique and valued attributes became firmly fixed, ensuring these precious breeds would stand the test of time.

Chapter 5.

Breeding, Bloodlines, and the Quest for Quality

October, 1904

Isabelle's Chicken Fricassee

Ingredients

Chicken .... 4 lbs	Water .... 1 Pint
Butter ....2 Tablespoonfuls	Grated Onion ... 1 Tablespoonful
Flour...............2 Tablespoonfuls	Salt, Pepper, Parsley

Directions

Select a cockerel which is fattened and not too old. Dress, singe and disjoint it. Put into a colander or frying basket and let the water run quickly over it. Wipe each piece dry. Put the butter into a cast iron pot and place over not too hot a fire; dip the pieces in flour. Put into the pot to brown on both sides. Remove the chicken and add the flour; stir well. Add the water and seasonings. When boiling, add the chicken. Cover and simmer gently for at least one hour or until tender. This must be cooked slowly or you will have tough and tasteless chicken. When ready to serve, arrange the chicken on a platter and strain the gravy over it. Garnish with triangles of toasted bread and parsley.

All domestic chickens are descendants of wild Junglefowl. All standardized breeds were created through multiple generations of selective breeding by humans. Ancient 'foundation breeds' such as Cochins, Dominiques, Dorkings, and Game Fowl were developed over centuries. They were fundamental in creating the composite breeds, including those four pioneer favourites: Plymouth Rocks, Orpingtons, Rhode Island Reds, and Wyandottes. Even the most outstanding breeds and strains needed continuous monitoring because, without meticulous human selection, bloodlines would degrade and gradually drift back toward their wild origins.

"Through careful and persistent breeding by expert poultrymen the American breeds mentioned elsewhere in this bulletin have been developed. In these new breeds the meat-producing qualities of Asiatic varieties and egg-producing qualities of European varieties are combined.

The knowledge of these facts should be of service to successful poultrymen, because of certain established laws in breeding. Like produces like, or a similarity of like; while another law, the law of revision is constantly at work. Remember that the present utility breeds trace back through the European and Asiatic breeds to the Jungle Fowl that laid only from 15 to 30 eggs in a season. Remember also that the law of reversion is counteracted only by careful selection year after year of the breeding stock having the desired type, characteristics and egg-producing qualities it is wished to perpetuate."

~ A.W. Foley, Poultry Superintendent for the Alberta Department of Agriculture, 1917

The General-Purpose Juggle

Throughout the decades, poultrymen and women have been caretakers of bloodlines, preserving breed identity and integrity through their selection and culling practices. Knowledgeable flock keepers mated their birds with care, using only superior specimens in breeding pens. Cockerels that did not measure up for breeding served another purpose; they were served up for dinner. But, selective breeding has always been a delicate balancing act. Too much emphasis on one trait could cause others to suffer. For example, flocks that are too heavily selected for egg production would begin to lose carcass quality, making them less valued for meat. Intense selection for feather patterns could lead to reduced egg production or lack of vitality.

Those who built strong, successful strains were expert jugglers, tossing several balls in the air simultaneously: phenotype, productivity, vigour, temperament, and all the other traits critical for the line's continued success. Respecting the original intent of the breed was another element in the equation, so finding that balanced-breeding sweet spot was an impressive achievement. Maintaining it through successive generations was an undeniable verification of one's skills. Bloodlines that retained and perpetuated a superior blend of breed characteristics became known for excellence and were widely distributed throughout North America.

Fine Lines

Just as successful companies crafted distinct brands to establish a strong identity in competitive markets, a similar trend emerged in the poultry world. A few high-performance strains gained widespread recognition. Some became household names, synonymous with reliability and consistently superior quality. According to I.K. Felch, a 'bloodline', or 'strain of blood' could be defined in a single word: 'uniformity'. The development of a strain, however, was not a process with hard and fast rules, and some breeders took shortcuts. Bloodlines were mixed haphazardly, released and marketed as a new strain before being fully established or proven. Felch addressed this practice in his 1902 book, Poultry Culture. "The American breeder is of a restless nature; he wants something that is peculiar to himself, something in which he can be identified. You find them all over the country chopping up the blood of their birds by the introduction of new sires, first from one flock, then from another, hoping thereby to have something different. They succeed; but when they have got it they are disappointed that no one else wants it. Now there is but one way to reach uniformity in breeding, no matter whether it is horses, cattle or fowls, and it is by 'in-breeding', and like poison, it may kill or cure, just according as we display good judgement in its use." Felch goes on to explain that a strain implies "strict adherence to the blood of a particular family, admitting no more foreign blood than is necessary to sustain the health and vigour of the race."

A Strain with Staying Power

In 1914, Clute & Walker had established themselves as a prominent name in western Canadian poultry circles. Their reputation was largely built on their strategic acquisition of the famed "Imperial Ringlet" strain of Barred Plymouth Rocks—an exceptional bloodline renowned for its superior conformation and crisp, parallel barring. Originally developed by E.B. Thompson of New York in the late 1800s, Imperial Ringlet Rocks gained widespread recognition throughout North America and beyond. Clute & Walker's success was deeply tied to the strain's proven qualities that permanently influenced the institution of poultry breeding.

First and Special Prize "Ringlet" Cock at New York, 1910-1911

Many fine males have good shape, color and barring, but this bird has more—
something not describable—something a great artist puts in his picture
that makes it the wonder of the ages—it is personality
life, spirit, style—a higher standard

Ringlett Barred Plymouth Rock

Arguably, the most prestigious poultry events of the late 1800s were the Madison Square Garden shows, sometimes called the 'Mecca of the Poultry Show World'. 1896 marked Thompson's eleventh consecutive year of exhibition successes at the annual event, with his now-famous Imperial Ringlets earning the highest acclaim.

Besides being showroom sensations, Ringlet Rocks boasted excellent utility traits such as egg size and production. However, exhibition awards and overflowing egg baskets were not the only measures of success for Thompson. He realized that a truly superior strain needed to shine in every way, and he went to great lengths to build strength and hardiness into his line. The Reliable Poultry Journal, a popular American magazine of the era, published some of the practices Thompson used to weed out weakness. He did not use incubators or brooders, and housed his birds in dozens of well-spaced houses covering 15 to 20 acres of pasture and meadowland. The hens raised their chicks in these 'brood coops' and ranged free with them throughout the spring, summer, and fall. "They are not nurtured and pampered. It is a case of survival of the fittest, with slight help. Only the more vigorous and hardy ones are desirable, either to breed from or to sell. Rigorous nature is permitted, to a certain extent, to weed out the unfit, the weakly ones."

Thompson recorded and studied the results of each mating, placing high value on individual birds that proved their ability to reproduce in high numbers. Touted as 'the greatest strain of Barred Rocks ever developed', the Ringlet name became legendary, a testament to the perpetuity of first-class breeding.

Standard Breed Supporters and Skeptics

The ever-increasing demands for food and livestock meant farmers prioritized productive fowl above all else. Opinions varied on the best methods to achieve this, and some outdated notions persisted. One such belief was that purebred, or 'standardbred' fowl were naturally less robust than crosses; a view reinforced by poultrymen such as William L. Allison in his 1885 American Standard Poultry Book.

> "I have done both myself; kept fowls for general use— ordinary common birds, mostly cross-bred—and kept pure-ly-bred birds to show, and I have no hesitation in saying that the former is the best plan, unless, of course, you are a poultry fancier and have money enough to indulge your mania for prize birds; then, with highly-bred stock, you may look to the sale of eggs and the taking of prizes to, in a measure, recoup you for your outlay.

> It is from experience then gained that I offer the following hints to those living in the country who wish to keep poultry and yet do not mean to incur much expense in so doing. For general use I would say do not keep entirely to pure-bred birds, but mix them with others; a good cross-breed is often more desirable than a really pure breed; not only are the fowls resulting from the cross stronger and less likely to become sickly and degenerate, but you can, by a judicious selection in the cross you allow, counteract many of the qualities you do not consider quite desirable."

The end of the nineteenth century brought pivotal changes to poultry breeding strategies, which helped settle some of the qualms around raising purebreds. Master poultrymen such as Isaac K. Felch were trusted advisors, having developed strains that bred true and stayed strong. Felch's breeding charts and tactics proved that a balanced approach to line breeding could, indeed, ensure the all-important utility traits of a standard-bred flock remained that way.

Breeders like Felch were also very attuned to the benefits of an outcross. They knew that when two birds "from widely different parentage" were mated, the progeny were more likely to be prolific and productive. To replicate and preserve what was known as 'outcross vigour', Felch used a percentage system of line breeding. His system of maintaining separate family groups guaranteed a continued, healthy population, without needing fresh infusions from outside. This process of gauging and controlling family re-latedness, combined with meticulous selection criteria, ensured a flock would remain stable, sound, and uniform.

By the early 1900s, it was a generally accepted principle that raising quality stock—whether four-legged or feathered—aligned with good farming practices and increased profits. Practical poultry keepers who had started with mongrels mostly shifted to a single breed. Still, the deleterious effects of close inbreeding remained a concern, and farmers were neither equipped for, nor invested enough, to implement a complicated breeding system. A simple and effective way to avoid inbreeding depression was to keep only the female offspring and regularly purchase replace-ment males. Once again, Alberta's leading poultry authority, A.W. Foley, cautioned against mixing breeds...or even bloodlines.

"If a variety of fowl has been bred pure for a great number of generations and in every generation only those allowed to live which conform to a certain type, the resulting offspring are very likely to possess a uniformity of type and characteristics similar to previous generations. A bird resulting from such a line of breeding is prepotent, which implies that it has the power to imprint its own likeness upon its offspring. If two birds, each of distinct lines of breeding are mated there is conflict. The characteristics of each strive to assert themselves with the result that one or the other may not be in evidence; or the offspring may

possess the characteristics of some remote ancestor. This
is not so true of the first cross as it is of the succeeding
ones, but, in any case, the benefit of the long line of careful
breeding is lost."

The Element of Exhibition

One other factor weighed into this complex composition of
breeding objectives: poultry exhibition. Success in the showroom
was both personally gratifying and bankable. The prospect of prizes,
money, and marketability incentivized exhibitors. Award-winning
birds brought accolades to the farm and attracted customers...
but not everyone came clamoring for exhibition breeding stock.
Producers recognized the economic value of a proven, dual-purpose breed, but many remained wary of 'exhibition' strains.

This lack of confidence was not a new development. It had arisen
during the mid- and late 1800s when two distinct poultry breeding
methodologies had taken hold in North America. Farmers/
producers favoured outbreeding (regularly introducing new
blood) to ensure a strong, productive flock. Exhibition breeders,
on the other hand, favoured line breeding. This controlled form
of inbreeding produced the more consistent offspring essential for
developing or fixing characteristics, building a strain, or excelling
in show halls.

Producers were less concerned with aesthetics, but fanciers
were driven to produce correct specimens—birds that adhered
closely to their standards. Exhibitors were accused of overem-
phasizing aspects of beauty: perfect combs, intricate plumage
patterns, or ornamental qualities such as crests, thereby allowing
utility functions to suffer. Selective pressures did vary somewhat
according to the desired outcomes, and the strengths most valued
by exhibitors did not always align with those of the producer.

Complex mating and selection strategies such as those introduced by Felch were designed to mitigate the negatives of line breeding, retaining both form and function in a strain. But the lines were drawn, and the utility vs exhibition dichotomy persisted. The importance (and even the possibility) of retaining uniformity in breeding, while preserving economic qualities, would remain an endless topic of debate within poultry circles for decades. A columnist for the Farm and Ranch Review offered his rather measured take on the issue in a 1914 opinion piece, "Are Purebreds Delicate?"

> "Close inbreeding is one of the curses of the hennery. The market poultryman, in order to have quick growing and hardy stock, changes his male birds every year or two. The fancier cannot well do that, or he will lose the good results of his matings. Yet the fancier, if he be a practical man, can so inbreed as not to be in danger. With all that, however, the less inbreeding done the better for the future generations. Mongrels are a product of a variety of bloods, and it is more difficult to inbreed them to any serious extent. They will not show it so quickly as a purebred, as the latter is bred more in line. Again, the average mongrel hen will not so readily overfatten for the reason that more or less Mediterranean blood exists in its make-up. It is only when we have birds of Asiatic crosses that we find a tendency to overfatten. There is no reason why a purebred should not be as hardy as a dunghill. And they will if they are not too closely inbred or fed too heavily on fattening food."

A quarter century later the dispute raged on, still at the forefront of poultry culture. Morley A. Jull, acclaimed author and head of the Poultry Science Department at the University of Maryland, tackled the contentious topic in his 1940 book, Poultry

Breeding. Jull maintained that breed and variety standards were part and parcel of the industry, and to lose sight of this would be a great sacrifice. He urged farmers to strive for the ideal combination that achieved both objectives. This, he said, was the only way to produce superior progeny and secure the best economic returns.

Substandard Stock

The craft of poultry breeding was, and remains, an incredibly challenging human endeavor with no guarantee of success. A high-performance bloodline could be maintained for decades in the hands of someone knowledgeable and competent, but the same line could be ruined in a generation or two with poor mating decisions. One thing is for sure...a bad experience leaves a bitter taste and a lasting impression.

Letter to the Editor, Farmer's Advocate and Home Journal, July 12, 1911:

Until eight years ago, most of my life was spent in large cities. I have always had the chicken fever. Whenever there was room for them on the city lot I had a few fancy fowls, and was much more interested in their show record than in the number of eggs they laid. When I came to Alberta to farm I firmly believed that I could produce high-class show birds and high record layers by the same line breeding methods. It cost me a good deal of money and wasted time to learn that I made a serious mistake.

On the start I had some White Wyandottes, purchased from a utility breeder who always out-bred; that is, he always purchased each year males not related to his hens.

They laid splendidly, much better than the scrub hens of my neighbours. I was very much pleased with the prospects for purebred poultry. Having made satisfactory beginning at a moderate cost with White Wyandottes, I sent to a famous breeder of this variety who had just won nearly all the prizes open to him in the leading poultry show of one of the states, and purchased a line bred trio, guaranteed to compete successfully with anything I might be required to meet. I shall never forget how elated I was when they came. Although I had paid a fancy price I was more than satisfied with my bargain. I have never seen a better cockerel in the poultry shows of Alberta. The pullet was as near perfect as one could ask for, and the hen was a good one, but during the entire year the hen laid but only eggs enough for three sittings, and the pullet did very little better. I had inquired particularly about the laying qualities of the strain while writing the breeder. He had assured me that there was no better, and spoke in high praise of the laying performance of the pullet which had just won first place at the state show. I have never written to a breeder of show birds concerning the matter and failed to receive a reply full of glowing prospects held out to the man who was after eggs.

It may be possible for a skilled breeder to line breed by using birds as distantly related as possible, and still be in the same strain and produce productive fowl, but it is certain to be a failure in the hands of a novice.

Last year I left the farm, where I had lived since coming to Alberta, and moved out to a claim. As I had no place to winter them, I sold all my fowls last fall. This spring I procured 24 Brown Leghorn pullets and young hens from a strain that has been outbred for years. They do not even

make a pretence at show color. I shall not say now many
eggs they lay, for I know a great many who read this would
not believe it. Two of them were not satisfactory, and I
killed them. There is a snappy, middle-aged Rhode Island
Red cockerel running with the remaining 22 females. He
is from a strictly utility strain, in no way whatever a show
bird. The eggs are as highly fertile as anyone could ask for.

~ W. I. Thomas, Alberta

Mr. Thomas was not alone in his disappointing experience
with a "show strain" which, in this case, resulted in his decision to
abandon purebred stock completely. Poultry specialists were aware
of occurrences like this one. They recognized that while some
producers had less-than-stellar results with a line, the advantages
of raising purebreds still outweighed the risks. Agricultural departments remained steadfast advocates for established, general-purpose breeds, and industry authorities urged producers to
maintain high-quality standards. This included prudent culling of
weak chicks, as described by the poultry department manager of
an Ontario experimental farm. "The weakling chick, the legacy
of the weak germ, will often peep itself to death—it may be in a
few hours, perhaps days from appearing well—and it is well that it
should do so; for he, or she who rears weakling stock is an enemy
to the best poultry interests of the country."

Women's Empowerment Through Poultry

A historical chronicle of poultry keeping would not be complete
without discussing the traditional roles of men and women in the
pre-WWI era, and how the pioneer culture loosened some of the
rigid Victorian rules of society.

Building a farm or ranching operation from a bare patch of land was, of course, gruelling work. By necessity, husbands and wives often worked as partners, distributing workloads according to individual strengths, abilities, and interests. There were many variations in how men and women divided their tasks, but some generalizations can be made.

Typically, pioneer men spent long hours labouring in the fields, growing and harvesting crops, and putting up hay. They were builders, fencers, woodcutters, and hunters. Men were generally the large livestock handlers and ranch hands. As soon as the boys in the family were deemed old enough to make a meaningful contribution, they began helping their fathers.

A day in the life of a pioneer woman was filled to the brim, caring for children and tackling domestic chores: baking, preparing meals, cleaning, and keeping up with sewing and laundry. Planting, growing, and harvesting vegetables was largely the woman's domain. Farm wives also helped with the crops, especially during the harvest season rush. They stooked grain, pitched hay, and drove teams of heavy horses. The lady of the house and her daughters shouldered many daily barnyard chores, such as milking the cow and feeding livestock. They separated cream, churned butter, and most women kept poultry.

In interviews and memoirs, farm wives and their daughters often spoke of long hours and staggering workloads. They also reminisced with pride in their accomplishments. The nature of the pioneer lifestyle afforded women some latitude to expand their skills beyond the kitchen, often blurring the strict conventions of the pervading patriarchal culture. The frontier life even empowered some women to enter the business world as commercial poultry producers.

Equality, however, would be a long time in coming. While women were caring for hens, raising chicks, and selling eggs, men were resoundingly recognized as the poultry experts. Poultrymen

Mrs. Herman Hungerbuhler and children, feeding poultry, near Vulcan, Alberta.", 1910

wrote articles on husbandry and breeding strategies. They became known for developing high-performance strains and competed with their birds in exhibitions. The knowledge they gained through experience and research was published, applauded, and respected.

Brushing aside the gender disparity, many Western women decided their little farmyard coops should be generating far better earnings than a small quantity of pin money. Those eager to participate more fully in the swiftly growing business of poultry production implemented effective breeding strategies to improve the performance of their flocks. They submitted written descriptions of their achievements to newspapers and shared their successes at Women's Institute meetings, encouraging and supporting each other.

Still, the North American breeding and exhibition sector remained cool and unwelcoming to the female population.

Women were acknowledged as flock caregivers, but their contributions were slow to be recognized in 'serious' poultry circles. Such ideologies did not discourage Mrs. Nettie Metcalf of Warren, Ohio. Nettie was a farm wife who spent years developing a new and distinct chicken breed. At first, her efforts to create a red, dual-purpose fowl were met with laughter from fellow fanciers, but as her new breed gained recognition, the ridicule turned to respect.

Nettie's Buckeye chickens were functional and meaty, praised for their foraging ability, hardiness, and gentle demeanor. The American Poultry Association gave the Buckeye breed its stamp of approval in 1905, proving beyond a doubt that poultry women were every bit as capable as men.

Pumping up Egg Production

Although the typical prairie farm family was not running an expansive poultry breeding operation, they knew the importance of selection, culling, and calculated mating choices. Improving egg production was a major objective at all levels of agriculture, from small chicken yards to provincial poultry plants. Table eggs remained a profitable sideline...especially for producers who managed to tap into those lucrative, but elusive, winter egg markets. Targeted mating strategies were an established pathway to improving a flock's performance, so retaining the best layers for use as breeding stock was essential. And this was only possible if the flock's top producers were identified. A Press Editor for the Farm and Ranch Review presented a simple method of finding the leading layers.

> "The pullet which will lay earlier than her mates hatched at the same time is apt to be more valuable as a layer than the slower laying birds, and should be marked with a leg band. It does not take long to band a pullet when she is

found on the nest, and make a record of her number; such a record will be found of great service during the year. When you see a hen on the nest, band her; if she is banded, take her number, and without making more trips than usual to the hen house, you can have a record which, though incomplete, will be of assistance."

This was not the most precise way to compile laying records, and those who demanded greater accuracy implemented a 'trap-nest' system. Trap nests were specially designed laying boxes with a hinged door that closed behind the hen. She would remain 'trapped' inside until released, allowing the egg-laying performance of individual hens to be captured accurately. The main drawback was the frequent trips to the henhouse for record keeping and releasing hens, ensuring an open box was available for the next occupant.

"The method I have formerly followed in securing my hatching eggs has been to pen up one or two lots of hens with selected males for the breeding season. I have never used any but purebred stock, and have always felt it necessary to cull severely to get my breeders. I have tried to get male birds each year from the same strain, but not too closely related to those used in the previous year. I find there is more even progress in improvement made thus than when varying the strain from year to year. Prepotency is just as likely to be lost by strain crosses within a breed as by crosses without, but, of course, it is not apparent in appearance and other visible qualities as much as it is in food provided.

I have always tried to use only hens for breeders, and as far as possible selected those that, as far as I could judge, were among the best producers in their pullet year. However, there is always guesswork about this. I know in my own case, and I believe I am more than ordinarily observant in all that pertains to feathered stock, I have made up my mind this is the last year I will breed from my stock without a trap-nest system.

In connection with this subject I have just been re-perusing Edward Brown's report on the poultry industry of Denmark, and he lays great stress on the effect trap-nesting, combined with a system of never breeding from any but two-year-old hens, has had on the average production in that country. The pullets are trap-nested their first year, and a small proportion of them is kept over for stock birds the next year. In one important point, Mr. Brown finds the Danes have surpassed both English and American breeders. They have not only paid great attention to number, but also to size of eggs, and the result is that a very large proportion of their output runs 17 and 18 pounds to 120 eggs."

~ A.B. Smith, Editor, Farmers Advocate and Home Journal, Winnipeg. March 8, 1910

Building Better Broilers

The economics of egg-selling was a convincing reason to raise the highly productive Mediterranean breeds, but specializing came with a downside: lighter-weight chickens did not produce a substantial carcass. A Leghorn cockerel could certainly become a meal, maybe a tasty chicken pie, but predominantly egg-laying breeds were not marketable as broilers, and along with all the other Canadian food commodities, the demand for chicken meat was rising fast.

To be clear, the concept of specific broiler-type chickens did not yet exist in the pre-industrial poultry world, where the trusty, general-purpose birds reigned supreme. Instead, the term 'broiler' meant young stock, usually cockerels, weighing from 1.25 to 2.5 pounds. Growing them out longer resulted in 'roasters' weighing from three to four pounds each. So, these chickens were not at all like modern broilers, plump and round, with mounds of white breast meat. The cooking fowl hanging in a boom-town butcher shop looked quite different, indeed.

In 1917, an Alberta Agriculture poultry specialist remarked, "A visit to almost any store handling poultry will demonstrate that a large amount of the dressed poultry offered for sale is poorly fleshed and equally poorly dressed. This is not because the demand for poultry is small, but through ignorance of the best method of fattening and dressing birds."

As with the egg industry, dressed poultry production was deficient in both quality and quantity. Once again, it fell to Alberta's Ag Department to remedy the situation. The department came up with several recommendations for producers, with specific instructions around crating, fattening, and selecting the best-typed stock to raise. Crating systems were designed to keep birds less active and consuming optimal feed rations. When cockerels were not expending energy running about the yard all day, scrapping amongst themselves and chasing females, their rate of gain increased. They finished up plumper and more tender. The flesh was reported to stay whiter if birds were fattened on ground oats mixed with sour milk, skim milk, or buttermilk. Corn was not advised as a finishing feed since it could cause breast meat to take on an undesirable yellow tint.

Anyone thinking about jumping into broiler farming in a big way had a few obstacles to work out, starting with meeting the

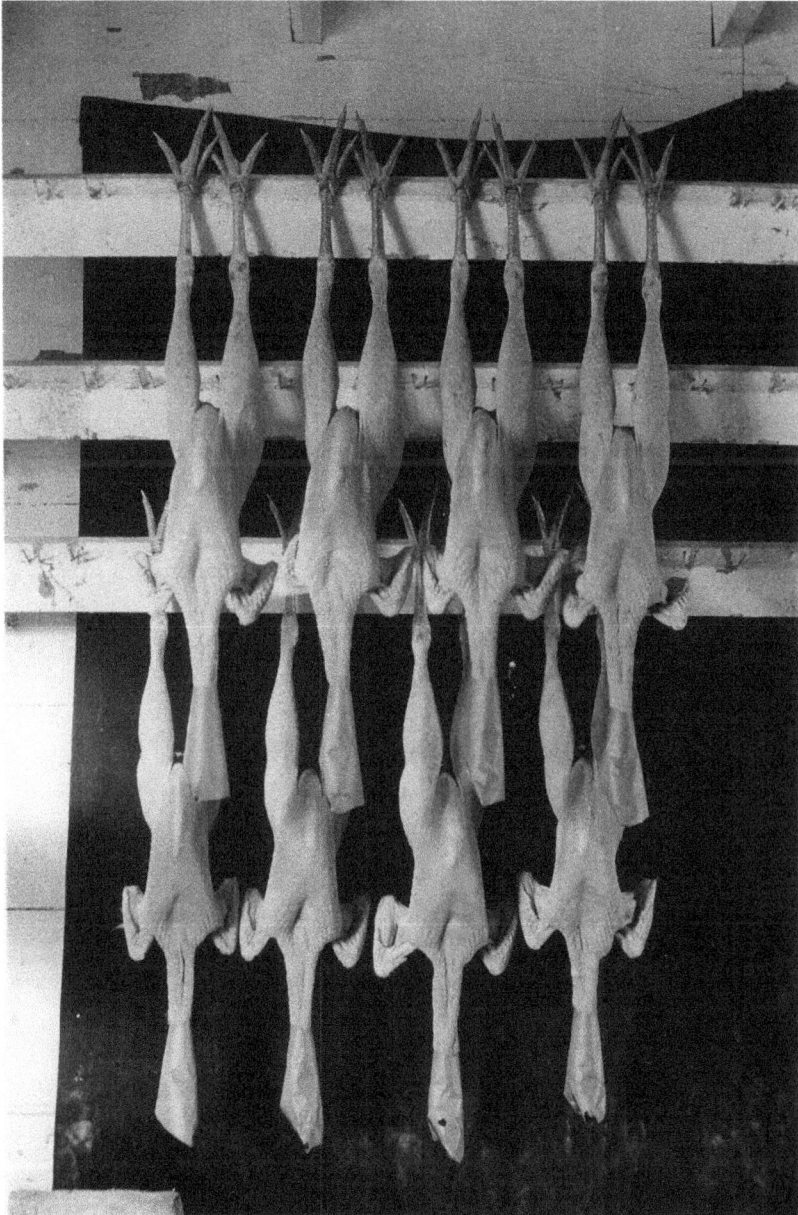

Rack of eight "naked" chickens ready for grading, University of Saskatchewan, University Archives and Special Collections, University Photograph Collection

demand for spring chicken. Sales were brisk from the middle of March to the middle of June, and to capitalize on that market, a producer would need to hatch chicks in the fall and winter. It would be necessary to purchase an incubator to turn out good numbers of chicks and deal with the extra work and risks of raising them during the coldest season. Plus, there was still that pesky problem of having a steady supply of fertile eggs at a difficult time of year.

A columnist for the Farm and Ranch Review got right to the meat of the matter in this 1914 article:

> "Broiler raising is a branch of the poultry business that calls for probably more patience and vigilance than any other. From the time the eggs are placed in the incubator, to the day the chicks are grown and marketed, there is constant care. Unless this close attention can be given, there will be considerable loss if not failure.
>
> The months of January and February are probably the best for hatching these chicks, as the stock can then be grown to marketable size in time to reach the top prices. The best eggs for broilers are those from a breed that has plump, fine grain meat, and birds that readily respond to good feeding. For this purpose our American breeds are best, and the leader of that class is the Wyandotte.
>
> During January and February, one hundred good, yearling hens should easily fill a machine capacity of from 100 to 200 eggs each week, and by having four of these small machines going—one started each week—there would be a new lot of chicks born every seven days, and after marketing began there would be a weekly shipment.

The business had better be started in a small way, so that whatever losses might occur would not be serious."

Getting agricultural products to city markets presented some challenges as well, but the Saskatchewan Department of Agriculture found a way to bring dressed chicken to Regina from the sparsely populated prairie producers. A 1915 Poultry Marketing Plan was an initiative of the Provincial College of Agriculture and the Canadian Northern Railroad. A baggage car was transformed into a mobile chicken processing station, operating on a six-week fall itinerary with stops in 34 towns. Farmers arrived with their wagon load of live chickens and were met with a skilled workforce, ready to help kill, pluck, and process the birds. Once they were graded, an advance payment was made to the farmer, and the birds were cooled, packed twelve to a box, and shipped off to Regina. The farmer would receive a final payment after the sale of his product, with the cost of boxes, transportation, and storage deducted.

Complexities of Capons

Caponizing, the surgical removal of the testes, was a practice that increased meat yields and enhanced the quality of male roasting fowl. The bird's finished weight could increase by as much as 20% by eliminating the hormones that would otherwise inhibit growth and fat accumulation. A capon also gained those pounds more efficiently than an intact male, and better feed conversion translated to increased profits for producers.

The meat of a capon was considered a delicacy; more tender, juicy, and mild tasting than the gamier roosters.

Caponizing the males brought flock management benefits, too. Aggression and territorial drive were greatly reduced or eliminated, allowing the big, mellow-mannered birds to be raised in large groups without harming one another. One manual even recommended using capons as 'nurse maids' for started chicks, stating they became as devoted to the task as a nurturing mother hen.

Despite these advantages, caponizing did not gain momentum among poultry farmers in Western Canada. Market commodity reports often included as many as 21 categories for dressed fowl, and another five or so for live, but capons remained distinctly absent. With the many benefits to be gained, it might seem surprising that more producers did not jump aboard the capon train. There was an obstacle standing in the way of raising capons, though, and it was a big one: surgery. The name "capon" comes from the Latin "capo," meaning "cut," and removing the testes from the body cavity was an invasive and sophisticated process.

The Farmer's Short Courses in Livestock outlined the procedure in detail. The age of the cockerel was of prime importance. Three to four months of age was the most appropriate time for the surgical procedure, when the body was roomy enough to operate on, but the testicles had not yet started to develop. The risk of bleeding to death greatly increased after the bird reached six months old due to the increase in blood supply to these organs. Specialized instruments, cleanliness, and expertise were the requisites for a successful surgery, but even the most proficient stockmen were bound to lose a few birds to bleeding or infection.

A common outcome, especially with beginners, was the creation of a 'slip.' A slip was neither a cockerel nor a capon, but somewhere in the middle and "possessing the mischievous disposition and the appearance of an ordinary cockerel, but, as a rule, being unable to reproduce." This condition could arise if even a tiny piece of testicle remained in the body.

In skillful hands, most caponized fowl healed quickly and could be turned out into an enclosed yard and fed soft food for a few days. After two weeks, the incision would be fully healed, and the fattening process could begin. Capons were normally grown longer than traditional roasting chickens, to about ten months old. Spring-hatched capons were not usually ready in time for Christmas markets but were processed and packed in January, February, and March.

The War Years

World War I took a tremendous toll on Canada. Over 650,000 young men and women left their homes to serve. More than 66,000 Canadians perished, and 172,000 were wounded over the war's four-year duration. The war years also had significant and lasting impacts on agriculture.

1914 marked the end of Canada's frontier boom and the start of the Great War. The phase of rapid settlement was disrupted, and the emphasis of agriculture shifted from supplying local markets to producing food for the war effort. Whereas earlier settlers had their hands full providing the next wave of newcomers with livestock and food, local consumer needs were now overshadowed. Food was urgently needed at the front, and by 1915, Ottawa was again pressuring the provinces to increase production.

The British Press joined the plea, advising Canadian poultrymen to expand their flocks and "prepare to profit from an unlimited demand for poultry and eggs". It was presented as an opportunity, but various factors made expansion impractical or impossible for most flock keepers. Wartime shortages caused grain prices to soar, driving up poultry feed costs. Those who had previously relied on purchasing feed grain for their flocks could not continue to do so and still turn a profit. Rather than growing their poultry businesses, many were forced to reduce or disperse their flocks.

The second factor was the loss of farm workers. Farmers were initially exempt from conscription since their work was vital to the war effort, but many farm boys felt compelled to join up anyway. Leaving their families to fight overseas caused farm labour shortages. Their chores could sometimes be redistributed to remaining family members, but the emotional impact of their absence must have been a much more difficult burden to bear.

As the war raged on, food production became a critical concern. Federal programs offered agricultural incentives and urged Canadians to maximize outputs as a matter of patriotism. The province took a more guarded approach, for poultry growers at least. Alberta's Department of Agriculture was formed in 1905 with a broad mandate that included supporting the folks who had colonized the province. A failed farming operation would benefit no one, so expansion needed to be tempered with caution.

In 1918, A.W. Foley, Poultry Superintendent, addressed the adverse farming conditions with reassurance that better times were ahead. Alberta poultrymen were advised, at the very least, to retain their best birds as foundation stock and rebuild productive flocks when the war ended.

Foley also offered words of encouragement to anyone in a position to increase their flock size, but emphasized this must not be misconstrued as a recommendation to start up a large-scale commercial operation. He explained how quickly the costs, equipment, and time requisites could spiral out of control. Such a venture, he explained, would require additional buildings, feed rooms, an icehouse, incubators, and perhaps even a labor-saving gasoline engine. Instead, a well-managed farmyard flock could bring a handsome profit with little investment. "So far as capital goes, the poultry industry affords great opportunities to every farmer—the capital required being time, careful attention to details, and an enthusiastic love for the work."

Initiatives to Instruct and Inspire

Right from the start, Alberta's Agricultural Department began developing stacks of reading materials on every aspect of the ag industry. Education was fundamental in growing the poultry sector, and pamphlets and booklets were printed and distributed, covering everything from coop design to egg marketing. Still, nothing builds skills like in-person instruction, and the western provinces came up with unique ways to bring learning opportunities to the people.

In one Alberta initiative, government workers took models of well-designed chicken houses around the province, stopping at fairs and other events throughout the summer. Department-operated poultry fattening stations were set up to demonstrate the advantages of crate-feeding systems. It was a successful plan, as the prototypes prompted several individuals to establish their own "fattening houses". Like a feedlot for cattle, the business model was

University of Saskatchewan, University Archives and Special Collections, University Photograph Collection, Better Farming Train

based on buying live cockerels, feeding them well, and reselling them as 'market-ready'.

You could say the Saskatchewan Better Farming Train was on the right track. The train was, in essence, a travelling agriculture school, bringing all the newest ideas and advancements to rural communities. From 1914 to 1922, it was full steam ahead as the Farm Train toured the province with its payload of modified train cars. At each stop, people could view exhibits, listen to lectures, and watch demonstrations on every aspect of rural living. A converted flat car became a stage for demonstrating "well-selected" horses, cattle, sheep, swine, and poultry. By 1922, over a quarter of the province's population had visited the Better Farming Train.

Alberta had its own version of locomotive learning, the "Mixed Farming Special". As the name suggests, on-train training covered a range of topics and livestock species, but the feathered kind was well-represented with live fowl displays, equipment, and exhibits of dressed poultry. The Ag Department's model chicken houses were also on board to make their rounds. In 1912, the Mixed Farming Special stopped at 60 locations around Alberta.

Demonstration Farms

The Dominion of Canada began operating experimental farm stations in 1886, and the scientific data they gathered proved highly beneficial to all areas of agriculture. Alberta took a page from their playbook and set up Provincial Demonstration Farms, each chosen to represent different climates and soil conditions. In 1908, the province established a facility in Edmonton dedicated entirely to the advancement of poultry.

The Provincial Poultry Station kept large flocks year-round, allowing them to trial different breeds, feeds, housing, and husbandry. Methods used in warmer climates, such as canvas-fronted hen houses, were determined to be insufficient for overwintering chickens in Alberta. Industries donated their newest incubators and brooders to the station for testing and research. Staff kept meticulous records, banding hens and tracking egg production. This research provided homesteaders with tools and information to find success raising chickens in this new land, while advancing the science of poultry-keeping as a whole.

Fairs and Exhibitions

Even before the North-West Territories were organized into provinces, breeders and producers gathered at agricultural fairs, drawn by prizes and the prestige of proving their stock was the best in the West. These events showcased agriculture and provided a perfect venue for government participation.

When Alberta joined Confederation in 1905, a Department of Agriculture was promptly formed. It hit the ground running with a host of pressing industry priorities to tackle. One such priority was to provide funding and management for fairs and exhibitions. This might seem a bit odd...a newly formed government agency placing such importance on fair collaboration, but the decision was guided by logic. Livestock exhibitions were recognized for their positive influences on producers, livestock, and rural communities. By 1909, the provincial government was involved in hosting four poultry show events in Alberta, which goes to show how deeply poultry exhibition is embedded in our rural landscape and culture.

Chapter 6.

Competition: a Catalyst for Growth

July, 1905

The sounds of laughter and chatter filled the air as Isabelle and Jonathan made their way through the crowded fairgrounds. Somewhere, a band struck up a lively tune; their fiddles and banjos created an infectious energy. The couple entered a long building filled with tables showcasing the season's bounty: pies and baked goods; bunches of slender carrots and round red beets; frilly-leafed Swiss chard, small prickly cucumbers, and red potatoes scrubbed clean. The rear wall of the building was hung with colourful patchwork quilts, a vibrant backdrop for gleaming jars of jams, jellies, pickles, and preserves.

They exited the bench show, squinting in the bright, midday sun. "Those pies looked delicious! Should we find some lunch?" asked Jonathan.

"I'm getting hungry too," Isabelle replied. "But listen, I hear crowing coming from that next building. Let's go see the poultry first!"

"As usual, chickens take precedence over my needs." Jonathan's smirk softened his words and elicited an eye roll from Isabelle. He guided her toward the open doorway, his hand resting gently on her back. They stepped inside to a cacophony of squawking, crowing, and quacking. A giant pair of African geese added their braying honks to the rustic symphony.

"I've never seen so many birds all in one place!" Isabelle was enthralled. "Oh! Look at that pen of Orpingtons! I could just bury my hands in all those golden feathers!"

Isabelle's chickens had been a blessing to her. She found contentment and comfort in caring for her flock; they had helped her transition to the often-lonely life of a frontier farm wife. Jonathan knew she missed her family dearly and the frequent outings intrinsic to her English upbringing. Their homestead held them captive with a heavy workload and animals to attend to, and this day was a much-needed break for them both. His spirits lifted in response to Isabelle's carefree joy. "By all means, let's go take a closer look. If you like them, maybe we should find the owner and see if he brought a cockerel here to sell. It would be a good chance to find a replacement for Prince."

"Well, let's make the most of this year's summer fair," Isabelle suggested, one hand resting on her rounded waist. "It's much easier to cart the baby around now than after it's born." Jonathan's wide smile spoke volumes. His dreams were all coming to fruition. He was a farmer, landowner, and soon-to-be father. He and Isabelle were creating a legacy for their child and, hopefully one day, grandchildren. His heart was full.

* * * * *

Strengthening the Sector through Exhibition

The summer or fall fair offered pioneer families a delightful break from everyday life. Town and country folks gathered together to cultivate a sense of community, even if they happened to be cheering for opposing teams in the tug-of-war or ball games! And no fair would have been complete without livestock shows. Poultry divisions were not included in every fair, but feathered farm stock competitions spanned the country at the turn of the century.

Poultry exhibitions were networking forums. They brought benefits to farmers, exhibitors, and the poultry sector at large. From the industry perspective, these events were perfect opportunities for grassroots training. Farmers came from miles around, eager to learn about the latest trends, breeds, and innovations. The expertise of poultry judges was also recognized as a valuable resource in the rapidly evolving dynamics of early 20th-century poultry keeping. In 1909, the Alberta Ag Department handed out grant money to 44 Agricultural Societies (that held fairs or exhibitions) and supplied judges to 35.

These were good reasons for producers to make the trip to a local show, even if they were not exhibiting. There were presentations and demonstrations, plus top-quality stock to admire. The judge's placings helped farmers learn what constituted the best specimens of a particular breed—practical knowledge they could take home and apply to their own flocks. They might also find value in comparing different breeds and varieties side by side. A display of well-fleshed Barred Plymouth Rocks might be the perfect visual aid for a farm wife to convince her husband it was time to ditch the 'dunghill' fowl currently occupying their yard. Of course, the show was also a social gathering where stories were told, experiences shared, and ideas exchanged.

For exhibitors, the benefits of showing birds were numerous. The major shows were stand-alone events, held separately from fairs. A poultry club usually organized its main show in the fall when the season's offspring were at or near maturity. For the serious poultryman, these shows were tantamount to business trips, prioritized and scheduled into their fall calendar.

Poultry shows were the breeder's window to the industry, and the top birds were a benchmark for quality. At a show, an exhibitor could compare his best birds with what others were raising—maybe find ways to up his game. A poultryman with a keen eye might identify qualities in a competitor's birds that could enhance his strain, and his new mission would be lining up a purchase. Feedback from a qualified judge was another bonus, as they collected fresh ideas from their travels around the country.

Of course, the ultimate goal of competing is winning, and having your bird claim a champion title would certainly be something to crow about! The thrill of victory could be a driving force on its own, but there were other, more tangible benefits. One was marketability. Top show awards brought public recognition and credibility to the exhibitor. Repeated wins were especially effective in validating breeding skills, and stock from a winning bloodline commanded premium prices.

The pursuit of awards and recognition caused many exhibitors to become extremely competitive. Seasoned poultrymen often guarded their secrets carefully, unwilling to share the hard-earned knowledge that gave them a competitive edge. Others were decidedly unwelcoming to novice exhibitors. In 1910, the editor of *Farmer's Advocate and Home Journal* urged breeders to set aside their personal ambitions in the interest of achieving a greater good. "The livestock breeder who, by his suggestions, helps another man to become an exhibitor of stock stimulates interest in the pure-bred livestock business, and even though he does increase the number of competitors in the show, directly helps himself and the breed he is interested in. The general diffusion of information on any phase of any agricultural industry is beneficial to the industry as a whole and profitable to everyone concerned. The more breeders who can be induced to the show, the better it will be for the shows, the better for the pure-bred stock business, and the better for the individuals concerned."

It is not known if this article had much effect in modifying behaviours, especially when considering the other major motivating factor: prize money. By 1918, the Alberta Agricultural Fairs Association had been formed to oversee the management of these events throughout the province. W. McAthey, Secretary for Viking Agricultural Society, spoke on the topic of prizes at an annual meeting of the Association. "To run a fair successfully," he stated, "is first of all to have an attractive prize list, one that will appeal to the small farmer as well as the large one, because if your exhibitors get the impression at all that the society is catering to one and not to the other you are not going to get the entries nor the attendance which naturally means a serious loss both ways financially." The Fair Association set entry fees at .50 per poultry entry and supplied prize money of $3.00 for class wins, such as the best pair of turkeys, geese, and ducks. Chicken class awards varied from show to show, often featuring specials for the most popular breeds in the region. Prizes were up for grabs in four chicken categories, according to a 1918 fair booklet:

Best Pair of Plymouth Rocks
Best Pair of Cochins, Brahmas, or Dorkings
Best Pair of Any Other Pure Breed
Pen of one Cock and Four Hens of Cross-bred Fowls

Western Canadian poultry show organizers needed to keep a finger on the pulse of exhibition and breeding trends, both locally and internationally. Poultry breeding was in an extremely dynamic, constantly evolving phase, and organizations needed to stay current on updates from the American Poultry Association, while acknowledging local trends.

Calgary's Annual Winter Show was to be held December 14-17, 1915, with a few changes announced to their 'usual classifications'. Cornish chickens were transferred from 'Games' to the English class. Blue Orpington and Blue Wyandotte exhibitors were encouraged to compete for prizes, even though Blue Orpingtons were still a few years shy of acceptance by the American Poultry Association. Proponents of Blue Wyandottes would have to wait another 60 years for recognition by the APA! But these showy varieties could still grace exhibition halls around the country and gain the adoration of spectators and fellow exhibitors.

Competition amongst Barred Rock exhibitors had become fierce. In hopes of capitalizing on breeder rivalries, Calgary Winter Show secretary announced some new sub-categories, specific to the variety. "The Barred Plymouth Rock breeders have been accorded as an experiment the following classes in addition to those heretofore given: Cockerel bred hen, cockerel bred pullet, pullet bred cockerel, and pullet bred cock. They claim this extra classification will produce a lot of entries."

Compact Competition

Bantam chickens were primarily raised for exhibition, and fanciers delighted in their highly competitive little show birds, taking pains to keep them in the finest condition. Historically, bantam show classes were arranged a bit differently than today. The 1905 APA Standard of Perfection outlined four classes that included Bantam chickens:

Moosejaw Poultry Exhibition, University of Saskatchewan, University Archives and Special Collections, University Photograph Collection

- Class VIII - Game Birds, both large fowl and bantam combined
- Class IX - Oriental Games: Indians (Cornish), Sumatras, Malays, and Malay Bantams
- Class X - Ornamental Bantams: Sebright, Rosecomb, Booted, Brahma, Cochin, Japanese, and Polish bantams
- Class XI – Miscellaneous: Silkies, Sultans, and Frizzles

Gaining a Competitive Edge

Ambitious exhibitors used a few strategies to improve their chances of taking home some prize money. Of course, high-quality birds formed the foundation of an exhibition line, but only a few would stand out and really shine in the spotlight. One important factor was their age.

W. Allison stated it best in his 1885 American Standard Poultry Book. "The Exhibitor who wishes to be early in the field must hatch early. Cockerels for August and September shows must be hatched out in January and February. In six to seven months the plumage of the cockerel is fairly perfected, and pullets, if kept from laying till six or seven months and then exhibited when laying has just commenced, will be in perfect order. Adult birds must be exhibited when over their moult, and when laying re-commences. The cock bird is not in spirits and will mope in his pen if shown before the moult is over, and the adult hen's comb will look dry and shrivelled up till she lays."

In selecting which birds to show, Allison discussed the importance of hatching plenty of chicks, removing inferior birds and gradually narrowing down to the top few specimens.

"A very large number of birds must be hatched from which to make a selection. Any with glaring disqualifications must be drafted out early in the season. A little later another lot must be cleared out, those with faults, but not sufficient to disqualify, leaving, say, twenty out of two hundred hatched. To these, give every possible advantage in the way of space and food. Some will answer expectations, and some will fail. Amateurs are often too sanguine, and imagine all are going to be prize-winners, whereas, it takes no little care and experience to attain the much-coveted honor."

The idea of a chicken competition isn't something new. And, it did not begin in show halls, with gentlemen strolling the aisles and scrutinizing the judge's choices with peevish grumbling. The earliest poultry competitions date back thousands of years and took place in a cockpit rather than a showroom. Cockfighting was a bloody sport, but it played a significant role in the initial spread and selective breeding of chickens. Some sources even state that cockfighting led to the domestication of poultry. Poultry exhibition has, for the most part, replaced the sport of cockfighting, but competition and poultry breeding share a history so long and intertwined that one could not have happened without the other.

Organized Leadership

In January 1873, fifteen prominent poultrymen convened in Boston for a historic meeting that laid the foundation for North America's oldest livestock organization: The American Poultry Association (APA). Mr. Wm. M. Churchman was appointed President. He ruled the roost along with five Vice Presidents, each selected from various states and from the province of Ontario. In response to the growing need for standardization of breeds and consistent judging criteria, the APA created North America's definitive guideline for poultry evaluation. Originally titled *The American Standard of Excellence*, this comprehensive guide detailed the physical characteristics of 46 chicken, duck, turkey, and goose breeds. Most breeds were further divided into varieties—distinguished by attributes such as plumage pattern, color, crest, or comb type—with meticulous descriptions of each.

This guidebook helped establish uniformity in poultry breeding and competition across North America, as judges were tasked with selecting the exhibits that adhered most closely to these written standards.

Both the provincial level of government and the Dominion of Canada viewed poultry judges as leading authorities in the field, trusting that their expertise would transform a competition into a learning hub for farmers. Poultry shows needed committees to run them, and the province pushed for the establishment of a provincial poultry organization. The concept became reality in 1913 when the Alberta Provincial Poultry Association was formed, with a membership that included all the industry leaders— judges, government officials, breeders, exhibitors, and producers. The role of the Association was to promote standard-bred poultry, educate the public, and organize exhibitions.

Show Reports with Substance

After a poultry show, a write-up would be submitted to agricultural newspapers and poultry publications. These summaries were usually much more than a simple list of winners; they were discussions of general breed strengths and deficiencies in the region. And some reports read like novels! The typesetters at the news office probably groaned at the stack of hand-written pages from a long-winded judge or show superintendent. Nevertheless, this was useful information, especially to the farmers and fanciers who had not made it to the show in person. Critiques and placings allowed readers to stay abreast of current trends. The observations drawn from the birds on exhibit represented a cross-section of what was being raised in the region. And, from the perspective of historical research, the highly detailed show report that follows is a timestamp; a moment captured in Western Canada's poultry history.

> Calgary Poultry Show – Farm and Ranch Review, December, 1912.
>
> "It is now becoming quite a task to write up a poultry show such as the late Calgary show. The days are gone by when we can just sit down and run off a short account, that might answer the purpose. In a short while we have jumped from an entry of 200 to one of 1,700, and what we now say about birds in a hurry may be dangerous if not carefully done, and some of us have not the time now to devote to such matters, i.e., just when they want doing. Therefore, if the report of the Calgary show is delayed, it is not the fault of the editor of the paper, but is the result of the tardy work of the judges.
>
> Probably very few, if any, even of the experienced poultrymen that took tickets to the Calgary show this year, expected anything like what was put before them, and it certainly was a great surprise to all of us that went in on the train to see the setting of the show, the quality for the birds and the numbers on view. The show was admirably benched in the big horse show building, and the first view was most inspiring and something quite beyond what most of us expected. The secretarial work was excellent, and everything moved along smoothly, and in order, so that we felt we were at a really up-to-date show, run on the right lines.

Asiatics – Light Brahmas were a nice lot, though we missed Mr. Wilson's birds and his enthusiastic and genial self. It is quite possible that some judges would have put the second cock first; quite a nice bird this, though the first, on this occasion is also a nice Brahma. Cochins were not remarkable as a lot, and do not require much analyzing. They lacked on the whole in size, and we missed the giants we have been used to seeing in the East.

Langshans were much nearer to tip top form, both in blacks and whites. Mr. Dewey annexed nearly all the principal prizes in blacks with a good lot, though another exhibitor pulled into first place with a very nice cockerel in that class. Type, as well as color, were distinctly in evidence, and the display was good. The whites, shown by Mr. Suitor, are away above the average, or above what one would expect. The two old cocks are on the right lines, and immense as regards size.

Barred Plymouth Rocks – the old cocks were not a very strong class, though the winner is a good sort, and won fairly easily. Hens were better on the whole. The winner is a very smart, good size, well barred and of the proper blue color. Cockerels were, on the whole, a good class, and showed considerable care in breeding, and we think, were mostly 'Alberta-bred.' The winner, except for being a bit plain in head, looks like making the best old cock when finished. He has more length of back, and more type than most of the others. There were several good cockerels, capital in barring but on the small side and too cobby in build. When full-grown, it was the opinion of the judge that they would not equal the first in size and type and Rock character, though other opinions might differ with this.

Pullets were a good class. The winner had size, markings and good Rock type, and rather easily defeated some of the smaller birds that depended on color alone. There was another good one in the same pen that the judge missed. The second was good, but was a little shafty – and there were others, several of which would have won other years, but could not this.

White Rocks were a disappointment, on the whole, after the showing of last year. The winning cock was nice and well-shown, the rest fair. Buff Rocks showed an advance, and so did some of the other colors.

Wyandottes showed considerable advance all along the line. The Golden-laced were more numerous and better than ever before. Silvers are also coming up. Buff, the best ever put down here, and a nice lot. Partidge want working up. There are too many pullet-breeding males shown as exhibition birds. We want black-breasts, thighs and fluff, not red or mottled on our exhibition males. Columbians came out much stronger than usual, in fact, this was the first real showing of the color.

The Whites were a big lot, and a good lot, quite the best lot yet put down, running strong all through.

Rhode Island Reds were heavy classes, and gave the judge, Mr. Ross, plenty of work to sort them out. There were many of good type and some of excellent color, especially in females. The young stuff showed much care in breeding and selection.

Games were fairly good, but not very numerous, the Indians making the strongest class, and, here, we may remark that the Indians were on the whole the right sort and well shown, condition was excellent.

Leghorns made strong classes all through and are coming up. The browns were a good lot, and would take up too much space to go over in detail. The males had color and type, and were well shown. In the females there was type and far better color that we have heretofore had to remark upon. In previous years we have been getting after the color of the brown females, but here there was marked improvement. S.C White cockerels were a good, big class and a strong class, for us out here and should do a lot of good to the breed. Buffs were better than usual. Blacks are not making much advance.

Anconas are making a gain and should do for they are a very useful breed. Campines were for the first time what we might call classes, and should be much more numerous later – all silvers.

Minorcas are still holding their own, though there was nothing very remarkable that was new. French are pulling up a little, and Houdans, especially, should do well in this climate for they are good, general-purpose fowls and hardy. Dorkings do not gain much, though those we have are nice specimens of the breed.

Orpingtons, Buff – A big lot and running strong all though. Blacks were good, and seem to have come to stay. The whites are picking up, but there is a good deal to be done here yet. The younger division were far better than the older ones, which is very encouraging. The winning cockerel and leading pullets will be heard of again. There seems to be a tendency to blue legs in this division, which will have to be watched.

Polands, especially White Crested Blacks, made a nice turnout. Hamburgs were useful; the biggest classes, the blacks. Silkies were quite in evidence, and there were some useful selling classes."

Some exhibitors went to great lengths to have their birds compete in every show possible. Even in the years leading up to the Great War, chickens, turkeys, geese, and ducks had become frequent railway passengers, traveling to exhibitions across Canada and south of the border. It was the start of the colourful and captivating era of the string men. These staunch poultrymen were experts at buying and selling poultry. They systematically travelled the show circuit, always on the lookout for promising birds that might win them some prize money, or make them a profit when sold further down the line.

Of course, most exhibitors could not leave their farms to accompany a feathered entourage for months at a time. But they could still enter their birds in a distant show and ship them there by rail. As noted in this report from an Edmonton exhibition in 1912, local exhibitors did not always welcome tough competition from afar.

EDMONTON POULTRY SHOW

This year the poultry exhibit was exceptionally good, being ahead in numbers of anything that has ever been held in Alberta.

The building – which is a fine one – is none too large for the number of entries, and no doubt by another year the management will see their way to give the poultrymen another building. I am sure it is a credit to the poultrymen of this province to have the President and Mayor acknowledge that more people visited the poultry building than any other building on the grounds.

This should encourage the poultrymen to go ahead till they have the finest show in Canada.

Quality was running high in almost all varieties. Everything was kept clean, the birds well fed and watered, and a great credit is due to the superintendent in charge, Mr. Nixon and his assistants, for the very able manner in which everything was attended to. If anything was unsatisfactory Mr. Nixon saw that it was put right immediately.

The judges were H. Ross, Calgary, and George Woods, Winnipeg.

The two cars of exhibition poultry brought in by Mr. Hoyt, Whitewater, Wis., and Mr. Warrington, Cornwall, Ont., made a large addition to the number of entries, which amounted to nearly 1300. There is a strong feeling amongst some of the exhibitors against allowing these to come in and take away a large amount of the prize money. This exhibition is open to the world, and I don't see how anyone can object to these people bringing in high-class birds and winning.

The poultrymen of Alberta should get down to business and get the best there is, and be willing to compete with anyone who wishes to exhibit, and let the best win. These outside exhibitors would not be given any advantage the local exhibitor does not get. They ought to be compelled to enter by the advertised date, and pay their entry fees at the same time (which I understand has not been the case), and thus put all the exhibitors on an equal footing, and this being done I see no reason why the Albertans should complain.

The Celebrated J. Shackleton

A handful of local individuals rose to celebrity status in the early 20th-century poultry world. Joseph Shackleton was one of these figures. His skills as a breeder, judge, and industry leader coalesced, earning him a reputation that spanned decades.

Joseph Shackleton homesteaded near Olds, Alberta, where he started gathering ribbons and awards under the banner of Mountain View Poultry Farms. He later moved to Edmonton and eventually retired in Vancouver, but not before he found great success on the prairies as a poultry breeder, exhibitor, and judge.

Joseph began making a name for himself at local fairs and poultry shows, often sweeping all the top prizes. At the 1910 Innisfail Fall Fair, Shackleton birds won first place in every one of the ten categories they competed in. Joseph was showing Brown Leghorns and Buff Orpingtons that year, as well as his soon-to-be-famous Shackleton Plymouth Rocks. Throughout the next decade, he racked up wins at major shows across Canada and into the United States. His Barred Plymouth Rocks broke records around the province, such as a 1912 gold medal win in Edmonton for Best Display in Show of Any Variety. Coming home from the Cleveland World Poultry Congress with a top award was a feather in his cap, rounding out a long list of achievements.

By 1919, Shackleton had slimmed down his poultry operation and raised Barred Rocks exclusively. The functionality of his strain set it apart, earning high acclaim in both exhibitions and egg-laying contests. Lethbridge hosted annual Dominion Egg Laying Contests, and in both 1919 and 1920, the winning pens were from Shackleton's strain. His birds or fertile eggs were available for purchase...but they were not cheap. Hatching eggs from his best pens were priced from $5.00 to $10.00 per setting.

Shackleton gained a solid reputation as a judge and was known for providing valuable feedback to exhibitors. This attribute was mentioned in an Edmonton Newspaper called the "Bulletin" in 1931. "Exhibitors at the Edmonton Poultry Show this year are particularly fortunate in having such well-known and able judges place the awards on their birds... Jos. Shackleton, of Vancouver, B.C., will officiate. The judges are always willing and anxious to help the exhibitors discern the good and bad points in their birds, and this is certainly a real educational feature of the show. Mr. Shackleton is well known to Edmonton and district exhibitors, being a former resident and exhibitor here himself, and a most able judge for a number of years."

While living in Edmonton, Shackleton worked with the Poultry Branch of Alberta's Department of Agriculture, managing the Poultry Plant and helping to produce educational materials. Somehow, he also found time to serve on the Executive Committee of the Alberta Provincial Poultry Association. In 1931, Shackleton was honoured at a Poultrymen's Dinner at the King Edward Hotel in Calgary. He was introduced to the gathering as an "outstanding poultry judge and Barred Rock specialist."

Honing a strain to excel in both egg production and exhibition is, perhaps, the ultimate breeding challenge. Shackleton's dedication and expertise left a lasting mark on prairie poultry culture, establishing a long-lasting legacy of excellence.

Counting Eggs

Competition is a powerful motivator. It pushed twentieth-century poultrymen to improve their skills, expand their facilities, and buy better stock. The competitive nature of poultry shows motivated breeders to fine-tune their lines to conform as closely as possible to the breed standards.

Some farmers and producers, however, felt exhibitions had little relevance to their work; it was a sport for fanciers only. Others realized there was merit in using show results to guide their purchases. Weight, structure, and muscling were key elements in both judging general-purpose fowl, and realizing profits in raising them, so the judge's evaluation of these traits was a strong indicator of marketability.

On the flip side, egg production had become a major consideration, and top show awards did not guarantee high egg yields. This is where egg-laying contests came in. Even people who did not recognize the benefits of an exhibition strain would view egg competitions as a reliable yardstick to measure quality. These contests provided quantifiable evidence of a strain's egg prolificacy.

Egg-laying contests not only benefited farmers; they generated valuable data on egg production across a range of breeds and strains. Each year, competitors looked for ways to increase yields and improve egg quality, experimenting with nutrition and housing. They used selective breeding to boost the laying capacities of their hens and extend seasonal laying cycles. These efforts led to huge advancements in the egg industry, helping it grow, thrive, and tackle the ever-increasing food demands of a growing population.

The Dominion Experimental Farm in Lethbridge, Alberta, held an annual Egg-Laying Contest. To compete, a farmer would send a pen of 10 hens to live at the Experimental Station for one year. This ensured an even playing field, with all hens receiving the same care, housing, and feed. Throughout the year, staff carefully tracked the egg-laying performance of each pen.

Alberta Egg-Laying Contest
Conducted by the Dominion Experimental Farm at Lethbridge, Alberta
Results for 52 weeks, from November, 1919, to October 31, 1920

Owner	Name of Breed...Total Eggs for Pen of 10 Birds
Northcott, W	Barred Rocks....................................1658
Ideal Poultry Yards	White Wyandottes 1477
Jones, S.H.	White Leghorns 1415
Hetherinton, A.E.	S.S. Hamburgs...............................1230
Young, E.H & Sons	White Leghorns 1200
Kerr & McGuinness	G. Campines....................................1188
Woods, Mrs. Jos	Barred Rocks.................................. 1143
Gregory, H.W.	White Wyandottes 1113
Timms & Eastwell	Black Wyandottes............................1025
Lockerbie, D.	White Wyandottes 965
Rhodes, Mrs. A.R.	Buff Orpingtons...............................950

The following year's contest began November 1st, 1920, and the early data pointed towards some different outcomes. This time, the Dominion Experimental Station included their own 'Dominion Barred Rocks' in the standings, a flock heavily influenced by the Joseph Shackleton strain. The contest manager was quick to point out that the Dominion hens were not actually competing in the contest or eligible for prizes; their outputs were being counted for comparison purposes only. It is a good thing they included that footnote as competitors would surely have had some strong objections. In the first 15 weeks, the Dominion flock was already far in the lead with 657 eggs recorded. Next in line was another pen of Barred Rocks with 496, followed by Single Comb

Anconas with 418, Single Comb Rhode Island Reds with 380, and Buff Orpingtons with 337. The best-performing flock of White Leghorns, a breed widely recognized as the queens of egg laying, was trailing in 18th position with just 114. It would have been interesting to learn the final, year-end results, but that information was, unfortunately, not found.

Egg competitions were an extremely effective catalyst for developing the egg industry. Producers became aware of the great variation in egg-laying capacities of hens, which meant there was vast potential for improving performance through selection and breeding. It was a formula that would be repeated a couple of decades later in the quest for better meat chickens.

By the end of the First World War, Western Canada's chicken and egg industry was still in its infancy, but winds of change were blowing across the prairies. Chicken farming was becoming profitable, artificial incubation technology was improving, and flock sizes were increasing. Electricity and gas engines would bring life-changing impacts to farming operations. New chicken breeds were under development, and more specialized approaches to poultry production would be explored.

The 1920s ushered in an era of growth and optimism, even as great economic hardships loomed on the horizon. But the poultrymen and women of the Western frontier were a hardy breed. From the humblest of beginnings, they had persevered, cultivating flocks, developing strains, and laying the foundation for a thriving industry. There would be opposition amongst poultry communities, conflicting ideas in breeding philosophies and practices. And there would be moments of unity, where producers, breeders, and exhibitors could find shared purpose and collaboration, drawn together in celebration of the function and beauty of a well-bred flock.

Kathryn Stevenson, May 20, 2025

Sources

Cover Photo Attribution:

Mabel Biggs feeding her chickens, Springfield Ranch, Beynon, Alberta.", [ca. 1910], (CU170662) by Biggs, H. B. Courtesy of Glenbow Library and Archives Collection, Libraries and Cultural Resources Digital Collections, University of Calgary.

Introduction

Stats on settlement and immigration
Selling the Prairie Good Life, Written by Graham Chandler, Posted September 7, 2016
https://www.canadashistory.ca/explore/settlement-immigration/selling-the-prairie-good-life
Image 100: Log Brooder House, Young Family Farm – Photo credit, K. Stevenson

Chapter 1 – A New Life in a New Land

John and Mary Walton Story - *Beaverlodge to the Rockies*, Beaverlodge and District Historical Association, Jan. 1, 1976
Betty (Euphemia) McNaught and Isabel Perry Interview, February 13, 1976. PAA ACC 81.279 #45
Mabel Barker, PAA interview, January 15, 1976. ACC 81.279 #20
The Biggs Legacy: https://en.wikipedia.org/wiki/Beynon,_Alberta

Coop Design: Foley, A.W. Poultry Superintendent, *Successful Poultry Raising,
Poultry Bulletin No. 3* Published by J.W. Jeffery, King's Printer, by Direction of
Hon. Duncan Marshall, Alberta Minister of Agriculture, 1918

Mrs. Cookson's Goose Story: Rollings-Magnusson, Sandra. *Heavy Burdens on
Small Shoulders*: pg 94; Wood, John Henry (1976 Autobiography. Accession
No. 76.45.PAA

Hungry Hawks: Avery Kenney Story: https://www.millard-and-kleinsteuber-
histories.com/

Chapter 1 Images

Image 101: "Mrs. Chapman washing clothes, Endiang, Alberta.", 1916,
(CU1127198) by Unknown. Courtesy of Glenbow Library and Archives
Collection, Libraries and Cultural Resources Digital Collections, University
of Calgary.

Image 102: Young Josephine Bedingfield on the family ranch, west of High
River, Alberta, 1915. Attribution: "Child feeding hens, Bedingfeld Ranch,
Pekisko, Alberta.", 1915, (CU1101331) by Unknown. Courtesy of Glenbow
Library and Archives Collection, Libraries and Cultural Resources Digital
Collections, University of Calgary.

Image 103: Woman Holding child in doorway of their log house on Brealy
Ranch, Big Hill Springs near Cochrane, Alberta in Winter. Attribution:
"Brealy Ranch, Big Hill Spring, Alberta.", [ca. 1900-1907], (CU1157447) by
Unknown. Courtesy of Glenbow Library and Archives Collection, Libraries
and Cultural Resources Digital Collections, University of Calgary.

Image 104: "Mabel Biggs feeding her chickens, Springfield Ranch, Beynon,
Alberta.", [ca. 1910], (CU170662) by Biggs, H. B. Courtesy of Glenbow
Library and Archives Collection, Libraries and Cultural Resources Digital
Collections, University of Calgary.

Chapter 2 – The Breeds That Built the West

Taggart, John and Allison, William L. *The American Standard Poultry Book*,
Copyright Crawford & Co. 1885

Poultry Breed Information:

https://livestockconservancy.org/heritage-breeds/heritage-breeds-list/
plymouth-rock-chicken/

https://americanbuckeyeclub.org/About_Us.html

Biggle, Jacob. *Biggle Poultry Book Number Three*, Biggle Farm Library, Wilmer
Atkinson Co. 1895

Stats on percentage of Alberta poultry consisting of chickens: Derry, Margaret
  Elsinor, *Art and Science in Breeding: Creating Better Chickens*, University of
  Toronto Press, 2012
Mary Edey Interview, PAA Interview, 14 April 1976 81.279 #56

Chapter 2 Images

Image 201: Silver Laced Wyandotte, Plate II Biggle Poultry Book
Image 202: Buff Orpingtons - F.L. Sewell, 1911, Reliable Poultry Journal
Image 203: Rhode Island Reds Print, Source Unknown
Image 204: Light Brahma, Biggle Poultry Book, 1895
Image 205: Bantams: page 117, plate XIII, Biggle Poultry Book
Image 206: Woman Feeding Turkeys, Peace River, Alberta. Peel's Prairie
  Postcard Library, Public Domain, https://www.library.ualberta.ca/peel
Image 207: Mrs. William Hawthorne feeding chickens and geese, Viking area,
  Alberta.
(ca. 1900-1903) Courtesy of Glenbow Library and Archives Collection, Libraries
  and Cultural Resources Digital Collections, University of Calgary.

Chapter 3 – Hatching, Brooding, Rearing

George and Pauline (Hamann) Betts, *Crestomere Sylvan Heights Heritage*,
  Published by Crestomere-Sylvan Heights Heritage. 1969
Swinging Coop - *Farmers Advocate and Home Journal*, May 20, 1907
H.H. MacPhie Letter - *Farmer's Advocate and Home Journal*, May 20, 1907
A. G. Gilbert Article - *Farmer's Advocate and Home Journal*, Winnipeg, , May 20,
  1907
Shipman Fireless Brooding Letter - *Farm and Ranch Review*, 1907
W.J. Bell Letter on Turkey Raising – *Farmer's Advocate and Home Journal*, May
  20, 1907
Poultry Pointers - *The Farmer's Short Courses in Livestock*, The F.B. Dickerson
  Company, Lincoln, Neb. 1921

Chapter 3 Images

Image 301: "Mrs. Roy Benson (Verna) and children with chicks, Benson
  homestead, Munson, Alberta.", 1912, (CU1128768) by Unknown. Courtesy of
  Glenbow Library and Archives Collection, Libraries and Cultural Resources
  Digital Collections, University of Calgary.

Image 302: "Tom Roycroft feeding poultry, Shanks Lake, Alberta.", [ca. 1915], (CU1117997) by Unknown. Courtesy of Glenbow Library and Archives Collection, Libraries and Cultural Resources Digital Collections, University of Calgary.

Image 303: Keystone Incubator and Brooder. *The American Standard Poultry Book*, Taggart, John and Allison, William L. Copyright Crawford & Co. 1885

Image 304: Turkey A Frame, *The Farmer's Short Courses in Livestock*, The F.B. Dickerson Company, Lincoln, Neb. 1921

Chapter 4 – From Pin Money to Profits

Egg Farming/Avoiding dirty eggs - *Farmer's Advocate and Home Journal*, December 20, 1911

Winter Egg Production and Egg Import Stats - Foley, A.W., Poultry Superintendent, *Successful Poultry Raising*, Poultry Bulletin No. 3, Published by J.W. Jeffery, King's Printer, by Direction of Hon. Duncan Marshall, Alberta Minister of Agriculture, 1918

Alberta Poultry Breeding Facility - Gerald Parlby, Department History 1905 – 1914. *Alberta Department of Agriculture Annual Reports* – The First Ten Years

Egg Circles for the Western Farmers – *Farm and Ranch Review*, Sept 5, 1912

Feather Products and Prices - *Farm and Ranch Review*, March 20, 1914

Hatching Egg Shipping - *Farmer's Advocate and Home Journal*, March 10, 1910

Mrs. Constantine Ad – *Farm and Ranch Review*, March 20, 1914

July Bargain Prices - *Farm and Ranch Review*, July 20, 1911

Role of the Producer - Derry, Margaret Elsinor, *Art and Science in Breeding: Creating Better Chickens*, University of Toronto Press, 2012

Chapter 4 Images

Image 401: Ad Section from *Farm and Ranch Review*, June 5, 1914

Image 402: PAA Photo #: A7693 "Manir Polet Feeding Chickens" (ca. 1907) Manir Polet feeding chickens on the farm near Villeneuve, Alberta. Public Domain.

Image 403: "Woman in costume, feeding hens on ranch in southern Alberta.", [ca. 1907-1908], (CU1115414) by Unknown. Courtesy of Glenbow Library and Archives Collection, Libraries and Cultural Resources Digital Collections, University of Calgary

Chapter 5- Breeding, Bloodlines, And the Quest for Quality

Chicken Fricassee Recipe – Kirk, Mrs. Alice Gitchell, *The People's Home Recipe Book*, Imperial Publishing Co. Toronto 1916

Definition of a strain - I.K. Felch, *Poultry Culture, How to Raise, Manage, Mate and Judge*, M.A. Donohue & Co. Chicago, 1902

Advice against mixing bloodlines - Foley, A.W. Poultry Superintendent, *Successful Poultry Raising, Poultry Bulletin No. 3* Published by J.W. Jeffery, King's Printer, by Direction of Hon. Duncan Marshall, Alberta Minister of Agriculture, 1918

Preserving Outcross Vigor – I.K. Felch, Poultry Culture, Part II, Chapter 1, Thoroughbred Fowls, M.A. Donohue & Co. Chicago, 1902

Outbreeding vs Line Breeding - Derry, Margaret Elsinor, *Art and Science in Breeding: Creating Better Chickens*, University of Toronto Press, 2012

Are Purebreds Delicate? - *Farm and Ranch Review*, March 20, 1914

Maintaining Breed Standards - Jull, Morley A., *Poultry Breeding*, John Wiley & Sons, Inc. Copyright 1940, Page 20

Nettie Metcalf's Buckeyes - https://americanbuckeyeclub.org/About_Us.html

Trapnesting - A.B. Smith, Editor, *Farmers Advocate and Home Journal, Winnipeg.* March 8, 1910

Broiler Farming - *Farm and Ranch Review*, March, 1914

Poultry Marketing Plan – *Farm and Ranch Review*, October 5, 1915

Caponizing - *The Farmer's Short Courses in Livestock*, The F.B. Dickerson Company, Lincoln, Neb. 1921

First World War Statistics - https://www.veterans.gc.ca/en/remembrance/wars-and-conflicts/first-world-war

Better Farming Train - https://en.wikipedia.org/wiki/Better_Farming_Train_(Saskatchewan)

Mixed Farming Special - Gerald Parlby, Department History 1905 – 1914. Alberta Department of Agriculture Annual Reports – The First Ten Years

Chapter 5 Images

Image 501: Ringlett Barred Plymouth Rock Photo – Artist F. Sewell, Public Domain

Image 502: Rack of eight "naked" chickens ready for grading, Filename a-3920c3_141.jpg Courtesy of University of Saskatchewan, University Archives and Special Collections, University Photograph Collection

Image 503: "Mrs. Herman Hungerbuhler and children, feeding poultry, near
Vulcan, Alberta.", 1910, (CU1108240) by Unknown. Courtesy of Glenbow
Library and Archives Collection, Libraries and Cultural Resources Digital
Collections, University of Calgary.
Image 504: Better Farming Train - Poultry Car: Filename A-1461_141.jpg
University of Saskatchewan, University Archives and Special Collections,
University Photograph Collection

Chapter 6 – Competition: A Catalyst for Growth

Government Support of Fairs - Gerald Parlby, Department History 1905 – 1914.
Alberta Department of Agriculture Annual Reports – The First Ten Years
https://open.alberta.ca/dataset/11410a8e-8825-48df-bc62-b843fa51e1e4/
resource/96f26019-cf23-483c-bab7-adb36f12f750/download/2836572-
1999-gleanings-annual-reports-department-agriculture-first-ten-years.pdf
1918 Poultry Classes – Annual Report of the Alberta Agricultural Fairs
Association for the Year 1918-1919, Published by the Association. PAA
73.307/175 Bx 14
1910 Bantam Classes - American Poultry Association, *The American Standard
of Perfection*, 1905 Edition, Published by the American Poultry Association,
Copyright 1905
Calgary Poultry Show Report – *Farm and Ranch Review*, December, 1912
Edmonton Poultry Show Report - *Farm and Ranch Review*, September 5, 1912
Lethbridge Egg Competition – *Directory of Poultry Breeders of Alberta*, Issued by
The Alberta Provincial Poultry Association, 1921. PAA 73.307/175 Bx 14

Chapter 6 Image

Image 601 – Poultry Exhibition, Moose Jaw, Courtesy of University of
Saskatchewan, University Archives and Special Collections, University
Photograph Collection. Exact date unknown, but prior to 1920.

Thank you for completing *Feathers on the Frontier*.

We would love if you could help by posting a review at your book retailer and on the PageMaster Publishing site. It only takes a minute and it would really help others by giving them an idea of your experience.

Thanks

PM Store Author's QR Code
https://pagemasterpublishing.ca/by/kathryn-stevenson/

To order more copies of this book, find books by other Canadian authors, or make inquiries about publishing your own book, contact PageMaster at:

PageMaster Publication Services Inc.
11340-120 Street, Edmonton, AB T5G 0W5
books@pagemaster.ca
780-425-9303

catalogue and e-commerce store
PageMasterPublishing.ca/Shop

www.ingramcontent.com/pod-product-compliance
Lightning Source LLC
Chambersburg PA
CBHW060350090426
42734CB00011B/2091